# Non-Equilibrium Thermodynamics for Engineers

**Second Edition**

T0344276

# Non-Equilibrium Thermodynamics for Engineers

## Second Edition

**S Kjelstrup**

*Norwegian University of Science and Technology, Norway*

**D Bedeaux**

*Norwegian University of Science and Technology, Norway*

**E Johannessen**

*Norwegian University of Science and Technology, Norway*

**J Gross**

*University of Stuttgart, Germany*

**World Scientific**

NEW JERSEY · LONDON · SINGAPORE · BEIJING · SHANGHAI · HONG KONG · TAIPEI · CHENNAI · TOKYO

*Published by*

World Scientific Publishing Co. Pte. Ltd.

5 Toh Tuck Link, Singapore 596224

*USA office:* 27 Warren Street, Suite 401-402, Hackensack, NJ 07601

*UK office:* 57 Shelton Street, Covent Garden, London WC2H 9HE

**British Library Cataloguing-in-Publication Data**
A catalogue record for this book is available from the British Library.

**NON-EQUILIBRIUM THERMODYNAMICS FOR ENGINEERS**
**Second Edition**

ISBN 978-981-3200-30-2

Desk Editor: Amanda Yun

Printed in Singapore

*This book is dedicated to our children*

# Preface of the Second Edition

The popularity of the first edition has promoted a revised and enlarged new print. Original chapters have been revised, especially the chapters on entropy production minimization. Two chapters concerning phase transitions and membrane transport (Chapters 9 and 10) have been added to take advantage of recent developments in surface transport. The eight lecture videos mentioned in the Preface of the First Edition are now available on iTunesU and MIT Open Academy (in collaboration with TU Delft.)[1] IUPAC notation has been introduced for symbol dimensions.

The authors were inspired by the Opinion paper issued on October 2015 by the Physical, Chemical and Mathematical Sciences Committee of Science Europe, D/2015/13.324/6, *A common scale for our common future*. The paper recommends policy makers to (quote) "guide the establishment of *exergy destruction footprints* for commodities and services". This, they argue, will enhance development of technologies with better energy efficiency.

In view of the impact such a development may have on climate issues, the teaching of non-equilibrium thermodynamics can become central, or even important. We hope that this book can help universities and schools worldwide to establish a curriculum in the field.

---

[1] http://theopenacademy.com/content/signe-kjelstrup

As said in the Preface of our first edition, we continue to welcome comments and feedback from users to help improve this book.

Trondheim and Stuttgart, January 2016

Signe Kjelstrup                           Dick Bedeaux
signe.kjelstrup@chem.ntnu.no              dick.bedeaux@chem.ntnu.no

Eivind Johannessen                        Joachim Gross
eivjohannessen@gmail.com                  gross@itt.uni-stuttgart.de

# Preface of the First Edition

Meeting the entropy challenge is probably more central than the issue of providing sufficient power to the world. The entropy production, not the energy used, can measure our wastes and the efficiency of work, or the limit of our activity. This book introduces non-equilibrium thermodynamics to engineers, and discusses how the theory can be useful for typical engineering problems.

The book has been written after many years of teaching the subject at the Norwegian University of Science and Technology, Trondheim, Norway, and the Technical University of Delft, Delft, The Netherlands. Early versions of the book have been used for short courses at the International Center of Thermodynamics, Istanbul, Chalmers Technical University, Gothenburg, Helsinki Technical University and Pennsylvania State University.

This book can be used in Bachelor or Master study programs after a basic course in thermodynamics, or for self study in the industry. The book requires knowledge of basic thermodynamics corresponding to that given by Smith, van Ness and Abbott in *Introduction to Chemical Engineering*, or in Moran and Shapiro's, *Fundamentals of Engineering Thermodynamics*.

To facilitate learning, exercises for the topics of the book and solutions to these, are available on the NTNU homepage.[2] Eight DVD lectures are also available there or from The Technical University of Delft.[3]

---

[2]http://www.chem.ntnu.no/nonequilibrium-thermodynamics/
[3]http://collegerama.tudelft.nl/mediasite/Catalog/?cid=0cbe1b45-06c6-4d03-a692-92a6dad4711d

Financial support from the Research Council of Norway is acknowl-edged. The authors are grateful to Statoil ASA for the cover picture from Mongstad.

The authors welcome comments and suggestions that can improve future editions.

Trondheim and Stuttgart, March 2010

Signe Kjelstrup                              Dick Bedeaux
signe.kjelstrup@chem.ntnu.no                 dick.bedeaux@chem.ntnu.no

Eivind Johannessen                           Joachim Gross
eijoh@statoil.com                            gross@itt.uni-stuttgart.de

# About the Authors

**Signe Kjelstrup** is Professor of Physical Chemistry since 1985 at the Norwegian University of Science and Technology (NTNU), Trondheim, Norway. Until 2015, she was also a part-time Chair on irreversible thermodynamics and sustainable processes at the Technical University of Delft, The Netherlands. Her works in irreversible thermodynamics concern electrochemical cells, membrane systems and entropy production minimization in process equipment. She holds an honorary doctorate from the University of North East China, and has been a guest professor at Kyoto University, Japan, University of Barcelona, Spain. Her book on irreversible thermodynamics, coauthored with K.S. Førland and T. Førland (Wiley, 1988 and 1994, Tapir 2001), has been translated into Japanese and Chinese.

**Dick Bedeaux** was Professor of Physical Chemistry at the University of Leiden, The Netherlands, from 1984 to 2002, and held (from 2002 to 2011) a part-time Chair at the Norwegian University of Science and Technology (NTNU), Trondheim, Norway. He is now emeritus at both places. Bedeaux, together with Albano and Mazur, extended the theory of irreversible thermodynamics to surfaces. He has worked on curved surfaces. Bedeaux is a fellow of the American Physical Society, and the recipient of the Onsager Medal from the Norwegian University of Science and Technology. Together with Jan Vlieger he wrote the book *Optical Properties of Surfaces* (Imperial College Press, 2002, and revised edition 2004).

**Eivind Johannessen** holds a Dr-Ing from the Norwegian University of Science and Technology (NTNU), Trondheim, Norway, and is presently a researcher at the Norwegian Energy Company, Statoil. His doctoral thesis on the state of systems with minimum entropy production was awarded Best doctor thesis defended at Norwegian University of Technology and Science in 2004.

**Joachim Gross** is Professor of Thermodynamics and Thermal Process Engineering at the University of Stuttgart, Germany. His research interest is in Molecular Thermodynamics and the development of Fluid Theories. After receiving his PhD from the University of Berlin, Germany, he worked in the Conceptual Process Design group of the BASF AG in Ludwigshafen for 4 years. In 2004, he became Associate Professor at the Delft University of Technology, The Netherlands, in the Separation Technology group. In 2005, he was appointed Chair of Thermodynamics at the same university, before moving to Stuttgart in 2010.

# Contents

# Chapter 1

# Scope

*The aim of this book is to present the essence of non-equilibrium thermodynamics and its use for engineers. The field was established in 1931 and developed during the forties and fifties for transport in homogeneous phases. Applications of the theory are now increasing. Some perspectives on the applications are given after a brief introduction.*

Non-equilibrium thermodynamics describes transport processes in systems that are not in global equilibrium. The field resulted from efforts of many scientists to find a more explicit formulation of the second law of thermodynamics. This already started in 1856 with Thomson's studies of thermoelectricity, see [1]. Onsager is, however, counted as the founder of the field with his papers from 1931 [2, 3], see also [4], because these put earlier research by Thomson, Boltzmann, Nernst, Duhem, Jauman and Einstein into a systematic framework. Onsager was given the Nobel Prize in Chemistry in 1968 for this work.

The second law is formulated in terms of the entropy production $\sigma$. In Onsager's formulation, the entropy production is given by the product sum of so-called conjugate fluxes, $J_i$, and forces, $X_i$, in the system. The second law then becomes

$$\sigma = \sum_i J_i X_i \geq 0 \qquad (1.1)$$

where $\sigma$ is larger than or equal to zero. Close to equilibrium each flux is a linear combination of all forces,

$$J_i = \sum_j L_{ij} X_j \tag{1.2}$$

and the reciprocal relations

$$L_{ji} = L_{ij} \tag{1.3}$$

apply. They now bear Onsager's name. In order to use the theory, one first has to identify a complete set of extensive *independent* variables, $\alpha_i$. The resulting conjugate fluxes and forces are $J_i = d\alpha_i/dt$ and $X_i = (\partial S/\partial \alpha_i)_{\alpha_{j \neq i}}$, respectively. Here $t$ is the time and $S$ is the entropy of the system. The three equations above contain, then, all information on the non-equilibrium behavior of the system.

Following Onsager, a consistent theory of non-equilibrium processes in continuous systems was set up in the nineteen forties by Meixner [5, 6, 7, 8] and Prigogine [9]. They calculated the entropy production for a number of physical problems. Prigogine received the Nobel prize for his work on dissipative structures in systems that are not in equilibrium in 1977, and Mitchell, the year after; for his application of the driving force concept to transport processes in biology [10].

The most general description of non-equilibrium thermodynamics is still the 1962 monograph by de Groot and Mazur [11]; reprinted in 1985 [12]. Haase's book [13], also reprinted [14], contains many results for electrochemical systems and systems with temperature gradients. Katchalsky and Curran developed the theory for membrane transport and biophysical systems [15]. Their analysis was carried further by Caplan and Essig [16] and more recently by Demirel [17]. Førland and coworkers gave various applications in electrochemistry and biology, and they treated frost heave [18, 19]. Their book presented the theory in a way suitable for chemists. Newer books on equilibrium thermodynamics or statistical thermodynamics often include chapters on non-equilibrium thermodynamics, see e.g. [20]. In 1998, Kondepudi and Prigogine [21] presented an integrated approach of basic equilibrium and non-equilibrium thermodynamics. Jou *et al.* [22] published the second edition of their book on extended non-equilibrium thermodynamics and Öttinger gave a non-equilibrium description of the nonlinear regime [23].

Non-equilibrium thermodynamics is constantly being applied in new contexts. Fitts gave an early presentation of viscous phenomena [24]. Kuiken [25] has written the most general treatment of multi-component diffusion and rheology of colloidal systems. Rubi and coworkers [26, 27, 28] used the internal molecular degrees of freedom to explore the development within a system. We are now able to deal with the law of mass action [29] for chemical reactions within the framework of non-equilibrium thermodynamics [12] and shall do so in Chapter 7. Bedeaux and Mazur [30] extended the theory to quantum mechanical systems. Kjelstrup and Bedeaux [31] wrote a book dealing with transports into and across surfaces. Chapter 9 gives an introduction to this topic, and Chapter 10 provides examples of its use in membrane transport. These efforts have broadened the scope of the theory.

Chemical and mechanical engineering needs theories of transport in systems with gradients in pressure, concentration, and temperature, see Denbigh [32, 33]. In isotropic systems there is no coupling between tensorial (viscous) and vectorial (diffusional) phenomena, so the two classes can usually be dealt with separately [12]. We concentrate on isotropic systems in Chapters 4–7.

Simple vectorial transport laws have long worked well in engineering, but there is now an increased effort to be more precise. The need for more accurate flux equations in modeling [34] increases the need for non-equilibrium thermodynamics. The books by Taylor and Krishna [34], Cussler [35] and Demirel [17], which present Maxwell-Stefan's formulation of the flux equations, are important books in this context. Krishna and Wesselingh [36] and Kuiken [25] have shown that the coefficients in the Maxwell-Stefan equations are relatively well-behaved, by analyzing an impressive amount of experimental data.

Non-equilibrium thermodynamics is necessary for a precise description of all systems that exchange heat, mass and charge. A frequent assumption is that there is a continuity in the intensive variables on

the phase boundary. In Chapters 9 and 10, we show how this assumption can be lifted, taking the interface resistance into account.

There is a constant need in mechanical and chemical engineering to
design systems that waste less work [37, 38, 39]. Fossil energy sources,
as long as they last, lead to waste that may harm the environment.
Better and more efficient use of energy resources is therefore central.
It is then not good enough to only optimize the first law efficiency.
The second law has also to be taken into account. The entropy production $\sigma$ can be used to measure the efficiency of a technical process.
Through non-equilibrium thermodynamics, one can develop methods
to improve the second law efficiency. One purpose of the book is to
present such methods, see Chapter 11.

The process industry may, in a not too distant future, be obliged to
report not only on $CO_2$ emission, but also on annual exergy destructed or entropy produced. Science Europe has proposed to introduce
an exergy destruction footprint.[1] Engineering companies are making
first efforts to move in this direction, by including possibilities for
exergy calculations in their process-modeling software. The public
sector can enhance such a development, by giving benefits to those
who limit their entropy production. The engineering community at
large can develop further tools to accomplish this. Efforts in several
fields, like control theory [40], are essential.

Non-equilibrium thermodynamics is the only theory that can be used
to assess the second law efficiency in detail, or how valuable resources
are exploited. It is an aim of this book to contribute to this issue.
The final chapter, Chapter 11, gives rules for energy efficient designs
in engineering practice.

We give in Chapter 2 the *characteristics of non-equilibrium thermodynamics* and explain why it is an important field. We recall how
the entropy production of an industrial plant is related to the lost
work in a plant. In Chapter 3, we show how to derive *the entropy
production* for systems with diffusion, conduction and chemical reactions. The derivations follow de Groot and Mazur [12] and Førland

---

[1]Physical, Chemical and Mathematical Sciences Committee Opinion Paper,
October 2015, D/2015/13.324/6

*et al.* [19]. Chapter 4 presents examples of *flux equations* for coupled transport of heat, mass and charge. Chapter 6 deals with shear flow and 7 with chemical reactions. Examples are used to show how the different processes are coupled. In Chapter 8 we estimate the lost work in an industrial process, the Hall-Heroult process for aluminum electrolysis. The entropy production by charge transfer and by heat transfer are both large. Chapters 9 and 10 describe how to deal with interface transport. Examples are taken from phase transitions and membrane transport. This is essential for separation technology. Finally, in Chapter 11, we describe a method to minimize the entropy production in process equipment. The method is described in detail for ideal gas expansion and energy efficient heat exchange in chemical reactors and distillation columns.

# Chapter 2

# Why non-equilibrium thermodynamics?

*This chapter explains what non-equilibrium thermodynamics is in more detail, and how it adds to engineering fields.*

The most common industrial and living systems have transport of heat, mass, charge, or volume, in the presence or absence of chemical reactions. The process industry, the electrochemical industry, biological systems, as well as laboratory experiments; are all systems that are out of equilibrium. Equilibrium thermodynamics is not sufficient to describe these. There are six main reasons why non-equilibrium thermodynamics theory [12, 13, 19, 31] is needed.

The theory

- gives an accurate description of coupled transport processes.

- gives the same systematic basis to all transport processes.

- can be used to define experiments.

- quantifies produced entropy, lost work or destructed exergy.

- specifies an entropy balance to use in thermodynamic modeling.

- can be used to optimize the energy efficiency.

The statements will be briefly illustrated in this Chapter and more in-depth in the rest of the book. Designing systems which dissipate less energy, is becoming increasingly more important [38]. At the end of this Chapter we give the relation between the entropy production of an industrial plant and the lost work or the exergy destructed in the plant.

## 2.1 Simple flux equations

Accurate expressions for fluxes are required in engineering. In order to see immediately what non-equilibrium thermodynamics can add to the modeling of real systems, we compare simple flux equations to flux equations given by non-equilibrium thermodynamics.

The simplest descriptions of heat-, mass-, charge- and volume transport are the equations of Fourier, Fick, Ohm, Darcy and Newton. Fourier's law expresses the measurable heat flux in terms of the temperature gradient by:

$$J'_q = -\lambda \frac{\partial T}{\partial x} \tag{2.1}$$

where $\lambda$ is the thermal conductivity, $T$ is the absolute temperature, and $x$ is the direction of transport.[1] Fick's law gives the mass flux of one of the components in terms of the gradient of its molar concentration $c$:

$$J = -D \frac{\partial c}{\partial x} \tag{2.2}$$

where $D$ is the diffusion coefficient. Ohm's law gives the electric current in terms of the gradient of the electric potential:

$$j = -\kappa \frac{\partial \phi}{\partial x} \tag{2.3}$$

where $\kappa$ is the electrical conductivity, and $\phi$ is the electric potential. Darcy's law says that the volume flow $J_v$ in a tube is proportional to the pressure gradient $\partial p/\partial x$ via the coefficient $L_p$:

$$J_v = -L_p \frac{\partial p}{\partial x} \tag{2.4}$$

---

[1]Following IUPAC, vectors and tensors are printed bold. In the simple cases in this and some subsequent chapters, the forces and the fluxes are only along the $x$-axis. The relevant equations apply only to these components, which are therefore not printed bold. A partial derivative $\partial/\partial x$ is used when the variable differentiated also may depend on the time.

And, a laminar flow in the $x$-direction with velocity $\mathbf{v} = (v_x, 0, 0)$ and velocity component $v_x = v_x(y)$ obeys Newton's law of friction:

$$\Pi_{xy} = -\eta \frac{\partial v_x}{\partial y} \qquad (2.5)$$

where $\Pi_{xy}$ is the $xy$-element of the viscous pressure tensor $\mathbf{\Pi}$, and the proportionality constant, $\eta$, is the shear viscosity.

The fluxes in Eqs. (2.1)–(2.5) are all caused by one gradient, or one driving force (see next Section). Fick's law, for instance, says that there is no mass flux if there is no concentration gradient. We know from experiments that a temperature gradient and an electric potential gradient also can give rise to a mass flux. To neglect such effects can have severe consequences. When used at interfaces, such assumptions can even be in conflict with the laws of thermodynamics [31, 41].

Non-equilibrium thermodynamics generalizes Eqs. (2.1)–(2.5) by taking all driving forces into account. The theory gives a common basis to all transport equations, and show how they are connected. The basis is the second law of thermodynamics as expressed through Eqs. (1.1)–(1.3). This means that the hydrodynamics of viscous fluids, the theory of diffusion, heat conduction, and chemical reaction, all have a common systematic basis [12]. One purpose of the book is to present this.

## Exercise 2.1.1

*In a stationary state there is no accumulation of internal energy, mass or charge. This means that fluxes of heat, mass, and charge are independent of position. The derivative of the above equations with respect to $x$ are then zero. From the first equation, we have:*

$$\frac{d}{dx} \lambda \frac{dT}{dx} = 0 \qquad (2.6)$$

*Equations like these can be used to calculate the temperature, concentration, electric potential and pressure as a function of position, when their values on the boundaries of the system and $\lambda$, $D$, $\kappa$, $L_p$ and $\eta$ are known.[2]*

---

[2]As the variables now only depend on $x$ we use roman d's in the differentials.

*Calculate the temperature as a function of position between two walls separated by 10 cm. The walls are kept at constant temperature, 5 and 25 °C, respectively. Assume that the thermal conductivity is constant.*

- **Solution:** According to Eq. (2.6) $d^2T/dx^2 = 0$. The general solution of this equation is $T(x) = a + bx$. The constants $a$ and $b$ follow from the boundary conditions. We have $T(0) = 278$ K and $T(10) = 298$ K. It follows that $T(x) = (278 + 2x \ cm^{-1})$ K.n

## 2.2   Flux equations in non-equilibrium thermodynamics

Many natural and man-made processes are not adequately described by the simple flux equations given above. There are, for instance, always large fluxes of mass and heat that accompany charge transport in batteries and electrolysis cells. The resulting local cooling in electrolysis cells may lead to unwanted freezing of electrolyte. Electrical energy is frequently used to transport mass in biological systems. Large temperature gradients across space ships have been used to supply electric power to the ships. Salt concentration differences between river water and sea water can be used to generate electric power. Pure water can be generated from salt water by application of pressure gradients. In all these more or less randomly chosen examples, one needs transport equations that describe coupling between various fluxes. The flux equations, Eqs. (2.1)–(2.5) then become too simple.

Non-equilibrium thermodynamics prescribes coupling among fluxes. Coupling means that transport of mass will take place in a system, not only when the gradient in the chemical potential is different from zero, but also when there are gradients in temperature or electric potential. Coupling between fluxes describes precisely the phenomena mentioned above. The coupling rules depend on the properties of the system.

**Homogeneous systems**

In a homogeneous (bulk) system with transport of heat, mass, and electric charge, we shall learn in Chapters 3 and 4 that the linear relations for the fluxes of heat, mass and charge Eqs. (2.1)–(2.3) take the form

$$J'_q = L_{qq} \frac{\partial}{\partial x}\left(\frac{1}{T}\right) + L_{q\mu}\left(-\frac{1}{T}\frac{\partial \mu_T}{\partial x}\right) + L_{q\phi}\left(-\frac{1}{T}\frac{\partial \phi}{\partial x}\right) \qquad (2.7)$$

$$J = L_{\mu q} \frac{\partial}{\partial x}\left(\frac{1}{T}\right) + L_{\mu\mu}\left(-\frac{1}{T}\frac{\partial \mu_T}{\partial x}\right) + L_{\mu\phi}\left(-\frac{1}{T}\frac{\partial \phi}{\partial x}\right) \qquad (2.8)$$

$$j = L_{\phi q} \frac{\partial}{\partial x}\left(\frac{1}{T}\right) + L_{\phi\mu}\left(-\frac{1}{T}\frac{\partial \mu_T}{\partial x}\right) + L_{\phi\phi}\left(-\frac{1}{T}\frac{\partial \phi}{\partial x}\right) \qquad (2.9)$$

The forces of transport conjugate to the vectorial fluxes $J'_q$, $J$ and $j$ are the thermal force $\partial(1/T)/\partial x$, the chemical force $[-(1/T)(\partial \mu_T/\partial x)]$, and the electrical force $[-(1/T)(\partial \phi/\partial x)]$, respectively. The subscript $T$ of the chemical potential, $\mu_T$, indicates that the derivative should be taken keeping the temperature constant. We return to this point in Section 3.2.

Tensors of the same order interact through the $L$-coefficients. These are the so-called phenomenological coefficients or Onsager coefficients. They must be measured. The Onsager coefficients on the diagonal of the matrix can be related to $\lambda$, $D$, and $\kappa$. They are called *main coefficients*. The off-diagonal $L$-coefficients describe the coupling between the fluxes. They are therefore called *coupling coefficients*. Another common name is cross coefficients. According to Onsager, we have here three reciprocal relations or *Onsager relations* for the coupling coefficients:

$$L_{q\mu} = L_{\mu q}, \qquad L_{q\phi} = L_{\phi q}, \qquad L_{\phi\mu} = L_{\mu\phi} \qquad (2.10)$$

The Onsager relations simplify the system. They reduce the number of independent coefficients from nine to six. Coupling coefficients are small in some cases but large in others. We shall see that large coupling coefficients lead to a small entropy production. If it is difficult to measure $L_{ij}$, we can rather measure $L_{ji}$. To measure both, gives a good control.

In Chapter 4 we shall learn how to write equations like the ones above, using the entropy production that we derive in Chapter 3. Exercise 2.2.1 shows how non-equilibrium thermodynamics is useful for the interpretation of experiments.

Chapter 7 and Chapter 6 will introduce fluxes that are tensors or scalars. These do not couple to the vectorial fluxes.

## Heterogeneous systems. Systems with internal variables

We shall see in Chapters 9 and 10 how fluxes and conjugate forces can be found also in heterogeneous systems, a system made of homogeneous subsystems separated by surfaces [31]. A force across a surface will appear as discrete. Coupling coefficients refer to the whole surface or layer. The scalar component of a vectorial flux into the surface, may now couple to a chemical reaction.

It is also possible to use internal variables, and write flux equations for the mesoscale. This is necessary for chemical reactions in order to describe experimental observations, see Chapter 7. The flux equation is then highly non-linear in the driving force. These developments in non-equilibrium thermodynamics are relatively new.

## Exercise 2.2.1

*Find the electric current in terms of the time independent and constant electric field $E = -\mathrm{d}\phi/\mathrm{d}x$, using Eqs. (2.7)–(2.9) , in a system where there is no transport of heat and mass, $J_q' = 0, J = 0$.*

- **Solution:** It follows from Eqs. (2.7) and (2.8) that

$$\frac{L_{qq}}{T}\frac{\mathrm{d}T}{\mathrm{d}x} + L_{q\mu}\frac{\mathrm{d}\mu_T}{\mathrm{d}x} = L_{q\phi}E \quad \text{and} \quad \frac{L_{\mu q}}{T}\frac{\mathrm{d}T}{\mathrm{d}x} + L_{\mu\mu}\frac{\mathrm{d}\mu_T}{\mathrm{d}x} = L_{\mu\phi}E$$

Solving these equations, using the Onsager relations, one finds

$$\frac{1}{T}\frac{\mathrm{d}T}{\mathrm{d}x} = \frac{L_{q\phi}L_{\mu\mu} - L_{\mu\phi}L_{q\mu}}{L_{qq}L_{\mu\mu} - L_{q\mu}^2}E \quad \text{and} \quad \frac{\mathrm{d}\mu_T}{\mathrm{d}x} = \frac{L_{qq}L_{\mu\phi} - L_{\mu q}L_{q\phi}}{L_{qq}L_{\mu\mu} - L_{q\mu}^2}E$$

Substitution into Eq. (2.9) then gives

$$j = \frac{E}{T}\left[L_{\phi\phi} - L_{\phi q}\frac{L_{q\phi}L_{\mu\mu} - L_{\mu\phi}L_{q\mu}}{L_{qq}L_{\mu\mu} - L_{q\mu}^2} - L_{\phi\mu}\frac{L_{qq}L_{\mu\phi} - L_{\mu q}L_{q\phi}}{L_{qq}L_{\mu\mu} - L_{q\mu}^2}\right]$$

The exercise shows that the electric conductivity that is normally measured as the ratio of measured values of $j$ and $E$, is not given by

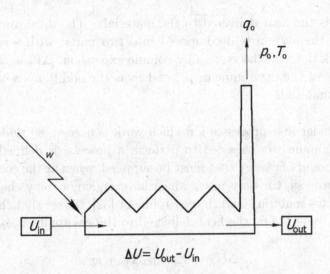

Figure 2.1: A schematic illustration of thermodynamic variables that are essential to find the lost work in an industrial plant.

$L_{\phi\phi}/T$ as one might have thought considering Eq. (2.9). The coupling coefficients lead to temperature and chemical potential gradients, which affect the electric current. The measured electric conductivity is the combination of the conductivity of a pure homogeneous conductor, found at zero chemical potential gradient and temperature gradients, minus additional terms. The combination of coefficients is an *effective* conductivity, or stationary state conductivity.

## 2.3   The lost work of an industrial plant

We shall see here why a second law analysis is important for the process industry. The industrial plant is a system, in which the materials undergo transformations. Materials are taken in and are leaving the plant at the conditions of the environment, see Fig. 2.1.

The environment is a good reference state for the thermodynamic analysis. It has constant pressure, $p_0$ (1 bar) and constant temperature, $T_0$ (for instance 298 K). We consider changes that take place during the time interval $\Delta t$. The first law of thermodynamics gives the change in internal energy of the process $\Delta U = U_{\text{out}} - U_{\text{in}}$:

$$\Delta U = q - p_0 \Delta V + w \qquad (2.11)$$

Here $q$ is the heat delivered to the materials. The total work delivered to the system is decomposed into two parts; with $-p_0\Delta V$ as the work done on the system by volume expansion, $\Delta V$, against the pressure of the environment, $p_0$, and $w$ as the additional work done on the materials.

We consider first processes for which work is needed, so that $w > 0$. The minimum work needed to perform a process, is defined as the least amount of energy that must be supplied, when at the conclusion of the process, the only parts which have undergone any change are the process materials and the environment [32]. To see that this is so, we first replace $q$ by the heat delivered to the environment, $q_0 = -q$:

$$\Delta U = -q_0 - p_0\Delta V + w \qquad (2.12)$$

In the classical formulation of the second law, we have

$$\Delta S + \Delta S_0 \geq 0 \qquad (2.13)$$

where $\Delta S$ is the entropy change of the process materials and $\Delta S_0$ is the entropy change in the environment. The total entropy change is positive. For a completely reversible process, the sum of the entropy changes is zero. For an irreversible process, the sum can be used to define the average total entropy production, $dS_{\text{irr}}/dt$, in the time interval $\Delta t$ as

$$\left(\frac{dS_{\text{irr}}}{dt}\right)\Delta t \equiv \Delta S + \Delta S_0 \qquad (2.14)$$

Assuming always that the environment behaves reversibly, the entropy change in the surroundings is $\Delta S_0 = q_0/T_0$. By introducing $q_0$ into the expression for the entropy production, and combining the result with the first law, we obtain

$$w = T_0\left(\frac{dS_{\text{irr}}}{dt}\right)\Delta t + \Delta U + p_0\Delta V - T_0\Delta S \qquad (2.15)$$

The left hand side of this equation is the work that is needed to accomplish the process with a particular value of the entropy production, $(dS_{\text{irr}}/dt)$. Since $(dS_{\text{irr}}/dt) \geq 0$, the *least* work requirement is

$$w_{\text{ideal}} = \Delta U + p_0\Delta V - T_0\Delta S \equiv E \qquad (2.16)$$

This is the work requirement for a completely reversible process. The equation defines the *exergy*, $E$, of the materials [38, 42, 43]. The exergy of the environment at $p_0, T_0$ (the reference state) is set to zero. We see that exergy analysis uses the classical formulation of the second law. It calculates the ideal work and the lost work (see below) from values at the system boundaries. Another word for exergy is availability [44]. Exergy can be used to measure the scarcity of minerals [43].

For processes that produce work, $w$ and $w_{\text{ideal}}$ are negative as defined above. Correspondingly, the signs of $\Delta U, \Delta V$, and $\Delta S$ are such that the system can perform work. The ideal work $w_{\text{ideal}}$ is minus the maximum work that can be obtained in the time interval $\Delta t$. This is the work performed by the system if the process is completely reversible. By comparing Eqs. (2.15) and (2.16), we see that $T_0 \left( \mathrm{d}S_{\text{irr}}/\mathrm{d}t \right) \Delta t$ is the additional quantity of work that must be used in the actual process, compared to the work in an ideal or reversible process i.e. $T_0 \left( \mathrm{d}S_{\text{irr}}/\mathrm{d}t \right) \Delta t$ is the so-called non-compensated heat of Clausius. By comparing Eqs. (2.15) and (2.16), we see that $T_0 \left( \mathrm{d}S_{\text{irr}}/\mathrm{d}t \right) \Delta t$ is the work that is lost in the actual process, compared to the available or useful work that can be produced in a reversible process. So, the lost work relative to the surroundings is:

$$w_{\text{lost}} = w - w_{\text{ideal}} = T_0 \left( \frac{\mathrm{d}S_{\text{irr}}}{\mathrm{d}t} \right) \Delta t \qquad (2.17)$$

This is the Gouy-Stodola theorem [45]. Equations (2.16)–(2.17) are central in the field that bears the name exergy analysis. In exergy analysis, the lost work or exergy destructed is calculated from knowledge of $w$ and $w_{\text{ideal}}$ in Eq. (2.17).

In non-equilibrium thermodynamics we can calculate the lost work by integrating the local entropy production, $\sigma$, over the volume of the system

$$\frac{\mathrm{d}S_{\text{irr}}}{\mathrm{d}t} = \int \sigma \mathrm{d}V \qquad (2.18)$$

The lost work is given by

$$w_{\text{lost}} = T_0 \frac{\mathrm{d}S_{\text{irr}}}{\mathrm{d}t} = T_0 \int \sigma \mathrm{d}V \qquad (2.19)$$

Non-equilibrium thermodynamics provides an explicit expression for $\sigma$ of Eq. (1.1) from Eq. (1.2). We have now two ways to find $dS_{irr}/dt$, from exergy analysis and from non-equilibrium thermodynamics.

The local resolution of $dS_{irr}/dt$ in terms of fluxes, forces and $\sigma$ is beyond the scope of exergy analysis, where a balance around the outside of a process (unit) is conducted. The explicit expression of $\sigma$ can be used to understand the origin of the entropy production or the lost work. At stationary state, the entropy production in a volume element is equal to the entropy flow into the surroundings of the element, see Section 2.5, and Chapter 3, Eq. (3.1). In Chapter 11 we show how a local resolution of irreversibilities can be used to optimize process units, by minimization of the entropy production.

**Exercise 2.3.1** *A cylinder of a combustion engine contains 600 cm³ of air at a pressure of 10 bar and a temperature of 1200 K, just before the exhaust valve opens (before the expansion starts). Determine the maximum available work (the exergy) of the air. Assume that air consists of ideal gases. The molar weight of air is $M_{air} = 28$ g.mol⁻¹. The temperature and pressure of the surroundings are $T_0 = 300$ K, $p_0 = 1$ bar.*

- **Solution:** The maximum available work per kg of gas is

$$|w_{\text{ideal}}| = u\left(T, p\right) - u\left(T_0, p_0\right) + p_0\left[v\left(T, p\right) - v\left(T_0, p_0\right)\right]$$
$$- T_0\left[s\left(T, p\right) - s\left(T_0, p_0\right)\right]$$

where the subscript 0 refers to values of the internal energy, the specific volume and the entropy of the air in the engine at the temperature and pressure of the surroundings. The thermodynamic functions were tabulated, see for instance Moran and Shapiro [46]

$$u\left(T, p\right) - u\left(T_0, p_0\right) = 719.3 \text{ kJ.kg}^{-1}$$
$$T_0\left[s\left(T, p\right) - s\left(T_0, p_0\right)\right] =$$
$$T_0\left[s(T, p_0) - s(T_0, p_0) + \frac{R}{M_{\text{air}}}\ln\frac{p_0}{p}\right] = 237.9 \text{ kJ.kg}^{-1}$$
$$p_0\left[v\left(T, p\right) - v\left(T_0, p_0\right)\right] = \frac{R}{M_{\text{air}}}\left(\frac{T}{p} - \frac{T_0}{p_0}\right) = -53.5 \text{ kJ.kg}^{-1}$$

This results in

$$|w_{\text{ideal}}| = 427.9 \text{ kJ.kg}^{-1}$$

for the maximum available work. Most of this work is normally not converted into useful work. The lost work is often significant in processes with chemical reactions, *cf.* Chapter 7.

**Exercise 2.3.2** *A heat reservoir has a temperature-independent heat capacity at constant volume $C_V$. The temperature of the environment is $T_0 = 300$ K. Determine the maximum available work (the exergy) of the heat reservoir when it has temperatures $T = 400, 4000$ and $40000$ K. The volume of the reservoir is constant, $V = V_0$.*

- **Solution:** The maximum available work is

$$|w_{\text{ideal}}| = U(T, V_0) - U(T_0, V_0) - T_0 [S(T, V_0) - S(T_0, V_0)]$$

We have:

$$U(T, V_0) - U(T_0, V_0) = C_V(T - T_0)$$

and

$$S(T, V_0) - S(T_0, V_0) = C_V \ln\left(\frac{T}{T_0}\right)$$

This results in

$$|w_{\text{ideal}}| = C_V \left[(T - T_0) - T_0 \ln\left(\frac{T}{T_0}\right)\right]$$

for the maximum available work. By introducing $T = 400, 4000$ and $40000$ K and $T_0 = 300$ K, $|w_{\text{ideal}}| = 13.7$ K $C_V$, $2923$ K $C_V$ and $38232$ K $C_V$, where $C_V$ has unit J.K$^{-1}$, respectively. The entropy contribution becomes negligible for $(T - T_0) \gg T_0$. The relative contribution of the entropy to the ideal work is larger, the lower is the temperature. We can thus expect larger losses of work at temperatures around $T_0$.

**Exercise 2.3.3** *Calculate the work available by mixing 1 mole of water into an excess of sea water. The water concentration in sea water is 54.9 kmol.m$^{-3}$, while it is 55.6 kmol.m$^{-3}$ in pure water. The temperature is 300 K.*

- **Solution:** In sea water we have $n_s$ moles of salt mixed with $n_w$ moles of water. The partial molar entropies of salt and water in seawater is $S_s$ and $S_w$ respectively. Thus the total entropy of the seawater is:

$$S_{sea} = n_s S_s + n_w S_w$$

In pure water the partial molar entropy of water is $S_w$. The total entropy of pure water and seawater before mixture is

$$S_{before} = n_s S_s + n_w S_w + 1.S_w^o$$

We have excess of seawater, so we assume that the partial molar entropies of salt and water do not change when only 1 mole of water is added. The total entropy after mixing is then

$$S_{after} = n_s S_s + (n_w + 1)S_w$$

The mixing entropy is

$$\Delta S_{mix} = S_{after} - S_{before} = S_w - S_w^o$$

For an ideal mixture the partial molar enthalpy of water is defined as

$$S_w = S_w^o - R\ln x_w$$

Here, $x_w$ is the mole fraction of water in seawater. The mole fraction is $54.9/55.6 = 0.987$. We have then that

$$\Delta S_{mix} = -R\ln x_w = 0.105 \ \text{J.(Kmol)}^{-1}$$
$$\Delta G_{mix} = -T\Delta S_{mix} = -32 \ \text{J.mol}^{-1}$$

Thus the available work from this process is 32 J.mol$^{-1}$ fresh water (*cf.* Section 10.5).

## 2.4   The second law efficiency. The exergy destruction footprint

In a work consuming process $w > w_{\text{ideal}}$, see Section 2.3. The second law efficiency is then:

$$\eta_{II} \equiv \frac{w_{\text{ideal}}}{w} = 1 - \frac{w_{\text{lost}}}{w} \tag{2.20}$$

This efficiency is also called the thermodynamic efficiency or the exergy efficiency. It includes $w_{lost}$. In a work producing process $w < w_{\text{ideal}}$ and

$$\eta_{II} \equiv \frac{|w|}{|w_{\text{ideal}}|} = 1 - \frac{w_{\text{lost}}}{|w_{\text{ideal}}|} \qquad (2.21)$$

The exergy destruction coefficient is $\xi \equiv 1 - \eta_{II}$. The definitions of $\eta_{II}$ in work (exergy) consuming or producing processes have values that vary between zero and one. An ideal reversible machine has $\eta_{II} = 1$, while a real machine has normally an efficiency far from one. A fuel cell, which is considered to be rather efficient, has typically $\eta_{II} = 0.6$. The efficiency refers to an unattainable reversible limit. An alternative practical limit is therefore proposed in Section 11.2; the state of minimum entropy production [47].

The Carnot process played an important role in the definition of the entropy by Clausius. The process starts with a volume of gas at pressure $p_A$ and temperature $T_h$. The system, in contact with a hot thermal reservoir with temperature $T_h$, is expanded isothermally to a pressure $p_B$. Subsequently it is expanded adiabatically ($q_{BC} = 0$) until it has the temperature $T_c$ of a cold thermal reservoir. The pressure has then been changed to $p_C$. The next step is to compress the system in contact with the cold reservoir at a constant temperature to a pressure $p_D$. Finally, the system is compressed adiabatically ($q_{DA} = 0$) to the original pressure $p_A$ and temperature $T_h$. The system has now returned to its original state, but an amount of heat $q_{AB}$ has been taken from the hot bath and converted into work $w_{\text{ideal}}$ and heat $|q_{CD}|$, added to the cold bath. In the reversible cycle of the Carnot process, the entropy production is zero, and the second law efficiency is unity. Stirling machines are examples of nearly reversible machines.

The first law efficiency [48] of a work producing process is defined as

$$\eta_I = \frac{w}{q} \qquad (2.22)$$

where $q$ is the heat added to the process. For the Carnot process, this efficiency depends only on the temperatures of the two reservoirs

$$\eta_I \equiv \frac{w_{\text{ideal}}}{q_{AB}} = \frac{q_{AB} + q_{CD}}{q_{AB}} = \frac{T_h - T_c}{T_h} \qquad (2.23)$$

where $T_h$ and $T_c$ are the temperatures of the hot and cold reservoirs, respectively. The first law efficiency is only close to unity if $T_h \gg T_c$. In a combustion process, the heat available for work is the enthalpy of reaction, giving $q = \Delta_r H$ and $w = \eta_I \Delta_r H$. This is no measure of how well the machine operates in terms of frictional and other losses. Such information can only be obtained from the second law efficiency, which measures how far the system is from reversible operation. The difference between $\eta_I$ and $\eta_{II}$ is elaborated for a plant producing aluminum, see Chapter 8.

More important than $\eta_{II}$ is to find all contributions to $dS_{\text{irr}}/dt$. When we multiply with the temperature of the surroundings, we have the exergy destructed by a process, or its exergy destruction footprint. Another name for the same is the lost work. The contributions give information on possibilities to operate or improve on the process, as explained in Chapter 8 and the last chapter on entropy production minimization. The introduction of exergy destruction footprints was proposed by Science Europe.[3]

**Exercise 2.4.1** *A saline power plant produces electric power from the mixing of sea water and river water to brackish water, at one bar and temperature $T_0 = 300$ K. One way to do this is by reverse electrodialysis, see Fig. 10.3. Discuss the second law efficiency of this plant.*

- **Solution:** The ideal electric work obtainable from an electro-chemical cell is given by Nernst's equation $w_{\text{ideal}} = FE = -\Delta G$ [48]. Here $\Delta G$ is the Gibbs energy difference of the mixing process, $E$ is the ideal cell potential and $F$ is Faraday's constant. We refer to one faraday of electrons transferred. The cell delivers in reality a smaller voltage, $E'$, due to its resistance and the concentration polarization inside the cell. The difference between $E$ and $E'$ is the lost work. The second law efficiency is

$$\eta_{II} = \frac{E'}{E}$$

---

[3]Opinion paper issued October 2015 by the Physical, Chemical and Mathematical Sciences Commitee of Science Europe, D/2015/13.324/6

The enthalpy of mixing $\Delta H$ is negligibly small for the mixing process. The ratio $-\Delta G/\Delta H$ obtained from $\eta_I$ is therefore clearly not a good measure of the plant's performance.

## 2.5  Consistent thermodynamic modeling

A thermodynamic model is a set of thermodynamic and other relations (equation of state, system variables etc., and flux-force equations) that are needed to solve the balance equations of the system. General balance equations are given in Appendix A.1.

Non-equilibrium thermodynamics offers possibilities to test the model for consistency. We mentioned a first test in Section 2.2. Two independent experiments can be done to find the coefficients $L_{ij}$ and $L_{ji}$. According to Onsager, these should be identical.

A second possibility is to use the entropy balance to evaluate the consistency of the thermodynamic model. The local entropy production can be calculated from Eq. (1.1) using the flux equations, Eq. (1.2), or alternatively, from the entropy balance (2.25). By integrating the local entropy production, $\sigma$, we obtain the total value:

$$\frac{\mathrm{d}S_{\mathrm{irr}}}{\mathrm{d}t} = \int \sigma \mathrm{d}V \qquad (2.24)$$

At stationary state, we also have from Eq. (3.1):

$$\frac{\mathrm{d}S_{\mathrm{irr}}}{\mathrm{d}t} = - \left( J_s^{\mathrm{i}} - J_s^{\mathrm{o}} \right) \Omega \qquad (2.25)$$

where $J_s^{\mathrm{i}}$ is the entropy flux into the volume, $J_s^{\mathrm{o}}$ the entropy flux out of the volume and $\Omega$ is the surface area through which the fluxes enter or leave the volume. The results should be the same.

Examples of entropy balances are given for heat exchangers, chemical reactors and distillation columns, see Tables 11.1 and 11.2. While $\mathrm{d}S_{\mathrm{irr}}/\mathrm{d}t$ depends on $L_{ij}$ and the local values of the thermodynamic variables, the entropy fluxes $J_s^{\mathrm{i}}$ and $J_s^{\mathrm{o}}$ can be calculated without knowledge of $L_{ij}$. When the entropy production from Eq. (2.25) agrees with the one found from Eq. (2.24), the model is consistent with the second law of thermodynamics.

Through such analyses one may reveal inconsistencies in assumptions that are made, either in thermodynamic relations or in the choice of parameters [49, 50]. To verify assumptions is essential for model improvements. Examples of consistency controls are given in Chapter 11, and Exercises 11.3.1 and 11.4.1.

# Chapter 3

# The entropy production of one-dimensional transport processes

*We derive the entropy production for a volume element of a homogeneous phase where diffusion, conduction, and chemical reaction can take place along the x-axis. The system is in mechanical equilibrium. Equivalent sets of pairs of conjugate fluxes and forces are derived.*

The second law of thermodynamics, Eq. (2.13), says that the entropy change of a system plus its surroundings is positive for irreversible processes and zero for reversible processes. This formulation of the law gives the direction of a process; it does not give its rate. Non-equilibrium thermodynamics assumes that the Gibbs equation remains valid locally. We shall see how the second law, as expressed by Eq. (1.1), results from Gibbs' equation. Rates of processes, as introduced through Eq. (1.2), will then have a thermodynamic basis. In this Chapter we consider systems where the transport processes are one-dimensional. They take place in the $x$-direction only. Many industrial applications concern one-dimensional diffusion and conduction. When there are surfaces, we consider transport only in the direction normal to the surfaces, *cf.* Chapters 9 and 10.

The change in the entropy in a volume element is the result of a flow of entropy into and out of a volume element, and of the entropy

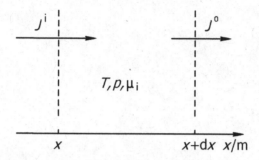

Figure 3.1: A volume element of a homogeneous phase with transport along the $x$−axis. $J^i$ and $J^o$ indicate fluxes that enter and leave the element. The element is in local equilibrium. $J$ indicates one of the fluxes.

production inside. The rate of change in the local entropy density is

$$\frac{\partial s}{\partial t} = -\frac{\partial}{\partial x} J_s + \sigma \qquad (3.1)$$

where $s$ is the entropy density per unit of volume, $J_s$ is the entropy flux and $\sigma$ the entropy production per unit of volume. By integrating Eq. (3.1) for stationary state conditions, we obtain Eq. (2.25). We shall now find an explicit expression for $\sigma$ by combining:

- mass balances,

- the first law of thermodynamics,

- the local form of Gibbs' equation.

We shall see that $\sigma$ can be written as the sum of the products of thermodynamic forces and fluxes in the system. These are the so-called *conjugate* fluxes and forces. The fluxes are used in subsequent Chapters to describe transport. The importance of $\sigma$ for determination and minimization of lost work, see Eq. (2.17), shall be dealt with in Chapters 8 and 11.

Consider a volume element between $x$ and $x+ dx$ of a container with an electroneutral homogeneous phase, see Fig. 3.1. The volume element does not move with respect to the walls of the container. It has

a sufficient number of particles to give a statistical basis for thermodynamic calculations. We assume local equilibrium in the element.
Its state is given by the temperature $T(x)$, the pressure $p(x)$, and the
chemical potentials $\mu_i(x)$. The system is in mechanical equilibrium.
This means that the system has no acceleration. The pressure in a
homogeneous phase is then constant. The assumption of local equilibrium is basic to irreversible thermodynamics. It has been tested
using non-equilibrium molecular dynamics and found valid for very
large temperature gradients [51]. Our aim is to find $\sigma$ in Eq. (3.1).

Symbols with units are listed at the back of the book.

## 3.1 Balance equations

The balance equations for the components of the system in the volume
element are

$$\frac{\partial c_j}{\partial t} = -\frac{\partial}{\partial x}J_j + \nu_j r \quad \text{for} \quad j = 1,...,n \quad (3.2)$$

where $J_j$ are the component fluxes, all directed along the $x$-axis, $\nu_j$
are the stoichiometric constants in a chemical reaction, and $r$ is its
rate in the volume element. The reaction Gibbs energy[1] is, see also
Chapter 7,

$$\Delta_r G = \sum_j \nu_j \mu_j \quad (3.3)$$

**Exercise 3.1.1** *Derive Eq. (3.1) by considering changes in a fixed
volume element.*

- **Solution:** The change of entropy is equal to the entropy flux
  into the volume element minus the flux out of the volume element, plus the increase in the entropy production. This gives

  $$\frac{\mathrm{d}S}{\mathrm{d}t} = -\Omega[J_s(x+\mathrm{d}x) - J_s(x)] + V\sigma$$

  The cross section is equal to the volume divided by $\mathrm{d}x$. We
  obtain in the limit of small $\mathrm{d}x$

  $$\frac{\mathrm{d}S}{\mathrm{d}t} = -V\frac{[J_s(x+\mathrm{d}x) - J_s(x)]}{\mathrm{d}x} + V\sigma = -V\frac{\partial J_s(x)}{\partial x} + V\sigma$$

---

[1]The driving force of the chemical reaction was called the affinity, $A$, by De
Donder, with $A = -\Delta_r G$, see e.g. [22].

By dividing this equation left and right by the volume, one obtains Eq. (3.1).

The conservation equation for charge is

$$\frac{\partial z}{\partial t} = -\frac{\partial}{\partial x} j \tag{3.4}$$

where $z$ is the charge density. The systems that we consider, can all be described as electroneutral. It follows that $\partial j / \partial x = 0$ so that the electric current density, $j$, is constant throughout the system.

According to the first law of thermodynamics, the change in internal energy is the net heat plus the work added to the system. For a change in the internal energy density $u = U/V$ per unit of time, we have for a volume element:

$$\frac{\partial u}{\partial t} = -\frac{\partial}{\partial x} J_q + Ej \tag{3.5}$$

where $J_q$ is the energy flux. In most of the cases in this book we shall use the definition

$$J_q = J_q' + \sum_{j=1}^{n} H_j J_j \tag{3.6}$$

The energy flux here is the sum of the measurable heat flux $J_q'(x)$ and the enthalpy flux carried by the component fluxes, $J_j$, where $H_j$ are the partial molar enthalpies. This definition has led to the name "total heat flux", which we shall use for this quantity throughout the book. For other definitions, see Appendix A.1 and Section 6.3.

The product $Ej$ in Eq. (3.5) is the electric power added to the volume element. The electric field, $E$, is often replaced by minus the gradient of the electric potential:

$$E = -\frac{\partial \phi}{\partial x} \tag{3.7}$$

The first law is illustrated in Fig. 3.2. Appendix A.1 gives a discussion of the relation between Eq. (3.5), the first law and the definition of the total heat flux.

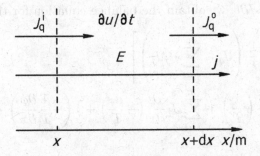

Figure 3.2: The internal energy change of a volume element with fluxes of total heat and charge across the boundaries.

## 3.2 Entropy production

The balance equations shall be combined with the Gibbs equation

$$dU = TdS - pdV + \sum_{j=1}^{n} \mu_j dN_j \qquad (3.8)$$

in order to find $\sigma$. We replace first the internal energy, $U$, entropy, $S$ and the mole numbers, $N_j$, with densities of the same variables, $u = U/V$, $s = S/V$ and $c_j = N_j/V$. By also using the fundamental relation $U = TS - pV + \Sigma_j \mu_j N_j$, we obtain the local form of Gibbs' equation

$$du = Tds + \sum_{j=1}^{n} \mu_j dc_j \qquad (3.9)$$

with $\mu_j$ as the chemical potential. The time derivative of the local entropy density is

$$\frac{\partial s}{\partial t} = \frac{1}{T} \frac{\partial u}{\partial t} - \frac{1}{T} \sum_{j=1}^{n} \mu_j \frac{\partial c_j}{\partial t} \qquad (3.10)$$

The use of partial derivatives indicates that the variables are also position dependent. By introducing the balance equations, Eqs. (3.2) and (3.5) into (3.10), using the rule for derivation of products, and solving for $\partial s/\partial t$, we obtain the balance equation for the entropy

$$\frac{\partial s}{\partial t} = -\frac{\partial}{\partial x}\left[\frac{1}{T}\left(J_q - \sum_{j=1}^{n}\mu_j J_j\right)\right]$$
$$+ J_q\frac{\partial}{\partial x}\left(\frac{1}{T}\right) + \sum_{j=1}^{n}J_j\frac{\partial}{\partial x}\left(-\frac{\mu_j}{T}\right) + j\left(-\frac{1}{T}\frac{\partial\phi}{\partial x}\right) + r\left(-\frac{\Delta_r G}{T}\right)$$

$$(3.11)$$

By comparing the above equation with the original Eq. (3.1), we identify the entropy flux in the system

$$J_s = \frac{1}{T}\left(J_q - \sum_{j=1}^{n}\mu_j J_j\right) = \frac{1}{T}J'_q + \sum_{j=1}^{n}S_j J_j \qquad (3.12)$$

and the entropy production

$$\sigma = J_q\left(\frac{\partial}{\partial x}\frac{1}{T}\right) + \sum_{j=1}^{n}J_j\left(-\frac{\partial}{\partial x}\frac{\mu_j}{T}\right) + j\left(-\frac{1}{T}\frac{\partial\phi}{\partial x}\right) + r\left(-\frac{\Delta_r G}{T}\right)$$

$$(3.13)$$

By replacing the total heat flux, $J_q$, with the entropy flux, $J_s$, we obtain an alternative expression

$$\sigma = J_s\left(-\frac{1}{T}\frac{\partial T}{\partial x}\right) + \sum_{j=1}^{n}J_j\left(-\frac{1}{T}\frac{\partial\mu_j}{\partial x}\right) + j\left(-\frac{1}{T}\frac{\partial\phi}{\partial x}\right) + r\left(-\frac{\Delta_r G}{T}\right)$$

$$(3.14)$$

We finally replace the total heat flux, $J_q$, with the more practical measurable heat flux, $J'_q$, by introducing Eq. (3.6) into Eq. (3.13). The result is yet another alternative expression:

$$\sigma = J'_q\left(\frac{\partial}{\partial x}\frac{1}{T}\right) + \sum_{j=1}^{n}J_j\left(-\frac{1}{T}\frac{\partial\mu_{j,T}}{\partial x}\right) + j\left(-\frac{1}{T}\frac{\partial\phi}{\partial x}\right) + r\left(-\frac{\Delta_r G}{T}\right)$$

$$(3.15)$$

where $\partial\mu_{j,T}/\partial x = \partial H_i/\partial x - T\partial S_i/\partial x$. The result was derived for electroneutral, homogeneous phases, without any other assumption than that of local equilibrium. Local equilibrium does not imply local *chemical* equilibrium!

**Remark 1** *Equation (3.15) shall be the starting point for many applications in this book.*

**Remark 2** *Equation (3.15) was derived from the Gibbs equation and the balance equations, and is valid whenever these equations are valid.*

**Remark 3** *The difference between the total heat flux and the measurable heat flux can be illustrated by an example. Consider a liquid and its vapor in a box. A constant heat flux, $J_q^{ll}$, is supplied to the liquid through the bottom of the box. A heat flux $J_q^{lg}$ is removed from the top of the vapor phase. From Eqs. (3.6) and (3.5) the energy flux is*

$$J_q = J_q^{l,l} + H_1^l J_1^l = J_q^{l,g} + H_1^g J_1^g$$

*In the stationary state, $J_1^l = J_1^g = J$, so*

$$J_q^{l,l} - J_q^{l,g} = (H_1^g - H_1^l)J$$

*The measurable heat fluxes in the liquid and the gas differ because of the source term at the interface. The total heat flux is constant. It cannot be measured in either of the phases, however, because $H_1^g$ and $H_1^l$ are not absolute.*

The entropy production contains pairs of fluxes and forces. We call the corresponding pairs *conjugate fluxes and forces*. In Eq. (3.15) the measurable heat flux $J_q'$ has the conjugate force $\partial(1/T)/\partial x$, the mass flux $J_j$ has the conjugate force $-\partial\mu_{j,T}/T\partial x$, the electric current density $j$ has the conjugate force $-\partial\phi/T\partial x)$ and the reaction rate $r$ has the conjugate force $-\Delta_r G/T$. The conjugate flux-force pairs in Eqs. (3.13), (3.14) and (3.15) are different. All these different choices are equivalent, however, and describe the same physical phenomena.

All force-flux pairs, except the last one, have a direction and are thus vectors. The reaction has a scalar flux-force pair, $r$ and $-\Delta_r G/T$. We are dealing with one-dimensional problems in this Chapter and all vectorial fluxes and forces are directed along the $x$-axis. A divergence of a flux reduces to the derivative of the $x$-component with respect to $x$. A gradient is similarly directed along the $x$-direction, and is given by the derivative of a potential or a temperature with respect to $x$. As it is known that, for instance a heat flux or an electric field, is vectorial, we do not complicate the notation further by making this distinction explicit.

Introduction of different fluxes, or changing the frame of reference, does not change the value or the physical interpretation of $\sigma$. The entropy production is an absolute quantity, as is the entropy. The preference for one of the expressions for a particular system, is always motivated by the system itself. We can illustrate this by an example: If the system is such that the chemical potentials of all components are constant, Eq. (3.14) is appropriate. In that expression the second term, containing the sum over $j$, is zero. One can of course also use the other two forms of the entropy production and simplify them using the constant nature of the chemical potentials. It is then easy to verify that the first and the second term combine and reduce to the first term in Eq. (3.14). Thus it is the properties of the system that determine which form of the entropy production is useful. We will use the last one, Eq. (3.15) when we describe experiments. The three alternative expressions are given as a help to find the appropriate form quickly.

The separate products do not necessarily represent pure losses of work (*cf.* Section 2.3). For instance, the electric power per unit of volume, $-\left(j\partial\phi/\partial x\right)$, does not necessarily give only an ohmic contribution to the entropy production, there may also be electric work included in the product, as we shall see in detail in Section 4.4. Each of the separate products normally contains work terms, and energy storage terms, see [19]. It is *their combination* which gives the entropy production and the work that is lost locally, $T_0\,\sigma$. The temperature of the surroundings, $T_0$, enters the formula for the lost work.

Dependent variables should be eliminated from $\sigma$. This simplifies the flux equations, and makes their solution easier. Such a simplification is one motivation for using electroneutral components in Eq. (3.2). Mass variables of this kind lead to straightforward reductions of Eq. (3.13). For instance, when the electric current density is zero, the electric power term disappears. The entropy production due to the chemical reaction disappears when the reaction rate vanishes, $r = 0$, or when the chemical reaction is in equilibrium, $\Delta_r G = 0$. Local chemical equilibrium introduces a relation between the remaining forces in the system. To see this, consider a reaction with $\Delta_r G = 0$. We can express the chemical potential $\mu_D$ of component D by the other chemical potentials contained in $\Delta_r G$. This can lead to a redefinition of the independent molar fluxes of the system. Gibbs-Duhem's equation (see also Exercise A.2.1 in Appendix A.2)

$$\mathrm{d}p = s\mathrm{d}T + \sum_{j=1}^{n} c_j \mathrm{d}\mu_j \qquad (3.16)$$

gives a further possibility for elimination of one of the chemical forces, $-(1/T)\partial\mu_j/\partial x$.

According to Prigogine's theorem [9] the expressions for $\sigma$, Eqs. (3.13)–(3.15), are valid in any frame of reference, which has a constant velocity v with respect to the laboratory frame of reference, provided that the system is in mechanical equilibrium. The mass fluxes change from $J_j$ in the laboratory frame to $J_j - c_j$v in a frame of reference with a constant velocity v, and the entropy flux changes to $J_s - s$v. The expression for the measurable heat flux remains the same. The transformation formula for the energy and entropy fluxes follow from those of the mass and measurable heat fluxes. We consider only systems that are in mechanical equilibrium, so Prigogine's theorem is applicable. A discussion and proof of the theorem was given by de Groot and Mazur [12], see also Chapter 6.

Haase [14] defined the dissipation function in non-equilibrium thermodynamics, analogous to the Rayleigh dissipation function for hydrodynamic flow, by $\Psi = T\sigma$. The integral over this function may be called D. An integration with $T > T_0$, gives $w_{lost} <$ D. The lost work was defined unambiguously in Chapter 2. It follows for cases with $T > T_0$, that parts of D can be extracted to do work. This means that D is ambiguous and leads to potentially incorrect results, when it comes to analysis of efficiencies in industrial processes. We shall avoid using $\Psi$ for this reason (see also [31] Section 4.2.1).

## 3.3 Examples

The exercises below illustrate in more detail how the contributions to the entropy production arise. The exercises 3.3.1–3.3.5 are meant to illustrate the theory. Exercises 3.3.6 and 3.3.7 give numerical insight, and 3.3.8 and 3.3.9 give physical insight.

**Exercise 3.3.1** *Consider the special case that only component number $j$ is transported. The densities of the other components, the internal energy, the molar volume and the polarization densities are all constant. Show that the entropy production is given by*

$$\sigma = -J_j \frac{\partial}{\partial x} \left( \frac{\mu_j}{T} \right)$$

- **Solution:** In this case the Gibbs equation, Eq. (3.10), reduces to

$$T \frac{\partial s}{\partial t} + \mu_j \frac{\partial c_j}{\partial t} = 0$$

The rate of change of the entropy is therefore given by

$$\frac{\partial s}{\partial t} = -\frac{\mu_j}{T} \frac{\partial c_j}{\partial t}$$

We use the balance equation for component $j$, Eq. (3.2), and obtain

$$\frac{\partial s}{\partial t} = \frac{\mu_j}{T} \frac{\partial}{\partial x} J_j = \frac{\partial}{\partial x} \left( \frac{\mu_i}{T} J_j \right) - J_j \frac{\partial}{\partial x} \frac{\mu_j}{T}$$

By comparing this equation with the entropy balance, Eq. (3.1), we can identify the entropy flux and the entropy production as

$$J_s = -\frac{\mu_j}{T} J_j \text{ and } \sigma = -J_j \frac{\partial}{\partial x} \left( \frac{\mu_j}{T} \right)$$

**Exercise 3.3.2** *Consider the case that only heat is transported. The molar densities, the molar volume and the polarization densities are all constant. Show that the entropy production is given by*

$$\sigma = J_q' \frac{\partial}{\partial x} \left( \frac{1}{T} \right)$$

- **Solution:** In this case the Gibbs equation, Eq. (3.10), reduces to

$$\frac{\partial u}{\partial t} = T \frac{\partial s}{\partial t}$$

The rate of change of the entropy is therefore given by

$$\frac{\partial s}{\partial t} = \frac{1}{T} \frac{\partial u}{\partial t}$$

The energy balance Eq. (3.5) reduces to

$$\frac{\partial u}{\partial t} = -\frac{\partial}{\partial x} J_q'$$

By substituting this into the equation above, we obtain

$$\frac{\partial s}{\partial t} = -\frac{1}{T}\frac{\partial}{\partial x} J_q' = -\frac{\partial}{\partial x}\left(\frac{1}{T}J_q'\right) + J_q'\frac{\partial}{\partial x}\left(\frac{1}{T}\right)$$

By comparing this equation with the entropy balance, Eq. (3.1), we can identify the entropy flux and the entropy production as

$$J_s = \frac{1}{T}J_q' \quad \text{and} \quad \sigma = J_q'\frac{\partial}{\partial x}\left(\frac{1}{T}\right)$$

## Exercise 3.3.3

*Consider a system with two components $(n = 2)$, having $\mathrm{d}T = 0$ and $\mathrm{d}p = 0$. Show, using Gibbs-Duhem's equation, that one may reduce the description in terms of two components to one with only one component.*

• **Solution:** Gibbs-Duhem's equation (3.16) gives

$$c_1\mathrm{d}\mu_{1,T} + c_2\mathrm{d}\mu_{2,T} = 0$$

The entropy production in Eq. (3.15) reduces for these conditions to

$$\sigma = -\frac{1}{T}\left(J_1 - \frac{c_1}{c_2}J_2\right)\frac{\partial\mu_{1,T}}{\partial x}$$

The equation contains only one (independent) force. Energy is lost by interdiffusion of the two components. We can also write this entropy production as

$$\sigma = J_V\left(-\frac{c_1}{T}\frac{\partial\mu_{1,T}}{\partial x}\right)$$

with $J_V \equiv J_1/c_1 - J_2/c_2$. This is the volumetric velocity flux of component 1 relative to the velocity, $J_2/c_2$, of component 2. Note that $J_V$ is independent of the frame of reference.

**Exercise 3.3.4** *Show how the chemical force in Eq. (3.15) can be derived from Eq. (3.13).*

- **Solution:** The chemical force in Eq. (3.15) can be rewritten as

$$\frac{\partial}{\partial x}\left(\frac{\mu_i}{T}\right) = \frac{1}{T}\frac{\partial}{\partial x}\mu_{i,T} + TS_i\left(\frac{1}{T^2}\frac{\partial T}{\partial x}\right) + \mu_i\frac{\partial}{\partial x}\frac{1}{T}$$

$$= \frac{1}{T}\frac{\partial \mu_{i,T}}{\partial x} + H_i\frac{\partial}{\partial x}\frac{1}{T}$$

By substituting this result into Eq. (3.13), we obtain

$$\sigma = \left[J_q - \sum_{i=1}^{n}H_iJ_i\right]\frac{\partial}{\partial x}\left(\frac{1}{T}\right) + \sum_{i=1}^{n}J_i\left(-\frac{1}{T}\frac{\partial}{\partial x}\mu_{i,T}\right)$$

$$+ j\left(-\frac{1}{T}\frac{\partial \phi}{\partial x}\right)$$

By using $J_q = J_q' + \sum_{i=1}^{n}H_iJ_i$, Eq. (3.15) follows. The new chemical force is related to the chemical potential by:

$$d\mu_{i,T} = d\mu_i - (\partial\mu_i/\partial T)_{p,\mu_j,j\neq i}\,dT = d\mu_i + S_i dT$$

**Exercise 3.3.5** *Derive the entropy production for an isothermal two-component system that does not transport charge. The solvent is the frame of reference for the fluxes.*

- **Solution:** In an isothermal system $\partial(1/T)/\partial x = 0$. Furthermore there is no charge transport so that $j = 0$. Finally the solvent is the frame of reference, so $J_{solvent} = 0$. There remains only one force-flux pair, namely for transport of solute. Using Eq. (3.15) we then find

$$\sigma = -\frac{J}{T}\frac{\partial \mu_T}{\partial x}$$

for the entropy production. Knowing that $\sigma \geq 0$ it follows that the solute will move from a higher to a lower value of its chemical potential.

**Exercise 3.3.6** *Find the stationary average entropy production due to the heat flux through a sidewalk pavement by a hot plate placed $d = 8$ cm under the pavement. The plate has a temperature of 343 K. The surface is in contact with melting ice (273 K). The Fourier type thermal conductivity of the pavement is 0.7 $W.K^{-1}.m^{-1}$.*

- **Solution:** Fourier's law for heat conduction is $J'_q = -\lambda(dT/dx)$. The entropy production per surface area is

$$\int_0^d \sigma dx = \int_0^d J'_q \frac{d}{dx}\left(\frac{1}{T}\right) dx = -\lambda \frac{\Delta T}{d}\left(\frac{1}{T_2} - \frac{1}{T_1}\right)$$

$$= -0.7\frac{(-70)}{(0.08)}\left(\frac{1}{273} - \frac{1}{343}\right) = 0.46\frac{W}{Km^2}$$

The lost work $w_{lost} = T_0\Omega \int_0^d \sigma dx$ per surface area $\Omega$, is $w_{lost}/\Omega$ = 275 K $\cdot$0.46 $W.K^{-1}.m^{-2} \doteq 125$ $W.m^{-2}$. It is typical for heat conduction around room temperature that the entropy production is large.

**Exercise 3.3.7** *In order to produce drinking water, one filters water through a 1 m thick sand layer with grain diameters around 0.1 mm. The height of the water column above the sand is $d = 1$ m, and the clean water outlet is at the top of the filter. Evaluate the stationary entropy production for a water flux of $10^{-6}$ $kg.m^{-2}.$ $s^{-1}$ at 293 K. The density of water is $\rho = 1000$ $kg.m^{-3}$, and the process can be considered to be in mechanical equilibrium.*

- The contribution to the chemical potential gradient is from the hydrostatic pressure gradient of the water column. The increase in the pressure of water at a distance $x$ from the sand surface is given by $dp = \rho g dx$. This gives $-d\mu_{w,T}/dx = V_w\rho g$, and

$$\sigma = J_w\frac{1}{T}V_w\rho g = 3.2 \times 10^{-8} W.K^{-1}.m^{-3}$$

This value is considerably smaller than the value for transport of heat to a pavement per $m^2$ surface (see exercise above).

**Exercise 3.3.8** *What is the entropy production for systems that are described by Eqs. (2.1), (2.2) and (2.3)?*

- **Solution:** Substitution of these equations into Eq. (3.15), setting the reaction rate zero, yields

$$\sigma = \frac{\lambda}{T^2}\left(\frac{\partial T}{\partial x}\right)^2 + \frac{D}{T}\frac{\partial \mu_T}{\partial c}\left(\frac{\partial c}{\partial x}\right)^2 + \frac{\kappa}{T}\left(\frac{\partial \phi}{\partial x}\right)^2$$

$$\equiv \sigma_T + \sigma_\mu + \sigma_\phi$$

Typical values in an electrolyte are: $\lambda = 2\,\mathrm{J.m^{-1}.s^{-1}.K^{-1}}$, $T = 300\,\mathrm{K}$, $\partial T/\partial x = 100\,\mathrm{K.m^{-1}}$, $D = 10^{-9}\,\mathrm{m^2.s^{-1}}$, $\partial \mu_T/\partial c = RT/c$, $c = 100\,\mathrm{kmol.m^{-3}}$, $\partial c/\partial x = 10^{-5}\,\mathrm{mol.m^{-4}}$, $\kappa = 400\,\mathrm{Si.m^{-1}}$, and $\partial \phi/\partial x = 10^{-2}\,\mathrm{V.m^{-1}}$. The resulting entropy productions are: $\sigma_T = 0.2\,\mathrm{J.K^{-1}.s^{-1}.m^{-3}}$, $\sigma_\mu = 10^{-13}\,\mathrm{J.K^{-1}.s^{-1}.m^{-3}}$ and $\sigma_\phi = 10^{-4}\,\mathrm{J.K^{-1}.s^{-1}.m^{-3}}$. Heat conduction clearly gives the largest contribution to the entropy production in electrolytes

**Exercise 3.3.9** *The Carnot machine converts heat into work in a reversible way. The efficiency is defined as the work output divided by the heat input [48, 46], see Section 2.4. This efficiency is $(T_h - T_c)/T_h$, where $T_h$ and $T_c$ are the temperatures of the hot and the cold reservoir, respectively. Compare this efficiency with the expression for the entropy production of a system that transports heat from a hot reservoir to the surroundings.*

- **Solution:** If we do not use the heat to produce work, but simply bring the hot and cold reservoirs in thermal contact with one another, there is a heat flow from the hot to the cold reservoir. The entropy production for the path, which has a cross section $\Omega$ is:

$$\frac{dS_{\mathrm{irr}}}{dt} = \Omega \int \sigma(x)dx = \Omega \int J_q'(x)\frac{\partial}{\partial x}\left(\frac{1}{T(x)}\right)dx$$

As there is no other transport of thermal energy, the heat flux is constant. This results in

$$\frac{dS_{\mathrm{irr}}}{dt} = J_q'\Omega \int \frac{\partial}{\partial x}\left(\frac{1}{T(x)}\right)dx = J_q'\Omega \int_{T_h}^{T_c}\frac{\partial}{\partial T}\left(\frac{1}{T}\right)dT$$

$$= J_q'\Omega\left(\frac{1}{T_c} - \frac{1}{T_h}\right) = J_q'\Omega\left(\frac{T_h - T_c}{T_hT_c}\right) = J_q'\Omega\frac{\eta_I}{T_c}$$

The work lost per unit of time, $T_c dS_{irr}/dt$, is thus identical to the work that can be obtained by a Carnot cycle, $\eta_I J_q' \Omega$, per unit of time. This can be regarded as a derivation of the first law efficiency of the Carnot machine. This machine is reversible and has as a consequence no lost work.

**Exercise 3.3.10** *Consider the reaction:*

$$B + C \rightleftarrows D$$

*The reaction Gibbs energy is:*

$$\Delta_r G = \mu_D - \mu_C - \mu_B$$

*In the absence of chemical equilibrium, the three chemical potentials are independent. The contribution to $\sigma$ from the reaction is:*

$$\sigma_{chem} = r \left( -\frac{\Delta_r G}{T} \right)$$

*Derive this expression for $\sigma_{chem}$, assuming that the reaction takes place in a reactor in which the internal energy is independent of the time.*

- **Solution:** The three balance equations for components B, C and D have a source term from the reaction rate, cf. Eq. (3.2)

$$\frac{\partial c_B}{\partial t} = -\frac{\partial}{\partial x} J_B - r$$

$$\frac{\partial c_C}{\partial t} = -\frac{\partial}{\partial x} J_C - r$$

$$\frac{\partial c_D}{\partial t} = -\frac{\partial}{\partial x} J_D + r$$

When the internal energy is independent of the time, the Gibbs equation, Eq. (3.10), reduces to

$$T\frac{\partial s}{\partial t} + \sum_{i=1}^{3} \mu_i \frac{\partial c_i}{\partial t} = 0$$

The rate of change of entropy is therefore given by

$$\frac{\partial s}{\partial t} = -\sum_{i=1}^{3} \frac{\mu_i}{T} \frac{\partial c_i}{\partial t}$$

We introduce the balance equations in this expression and obtain

$$\frac{\partial s}{\partial t} = \frac{\mu_B}{T}\left(\frac{\partial}{\partial x}J_B + r\right) + \frac{\mu_C}{T}\left(\frac{\partial}{\partial x}J_C + r\right) + \frac{\mu_D}{T}\left(\frac{\partial}{\partial x}J_D - r\right)$$

$$= \frac{\partial}{\partial x}\left(\frac{\mu_B}{T}J_B + \frac{\mu_C}{T}J_C + \frac{\mu_D}{T}J_D\right) - J_B\frac{\partial}{\partial x}\left(\frac{\mu_B}{T}\right)$$

$$- J_C\frac{\partial}{\partial x}\left(\frac{\mu_C}{T}\right) - J_D\frac{\partial}{\partial x}\left(\frac{\mu_D}{T}\right) + \left(\frac{\mu_B}{T} + \frac{\mu_C}{T} - \frac{\mu_D}{T}\right)r$$

By comparing this equation with Eq. (3.1), we can identify the entropy flux as

$$J_s = -\left[\frac{\mu_B}{T}J_B + \frac{\mu_C}{T}J_C + \frac{\mu_D}{T}J_D\right]$$

and the entropy production as

$$\sigma = -J_B\frac{\partial}{\partial x}\left(\frac{\mu_B}{T}\right) - J_C\frac{\partial}{\partial x}\left(\frac{\mu_C}{T}\right) - J_D\frac{\partial}{\partial x}\left(\frac{\mu_D}{T}\right)$$

$$+ r\left(\frac{\mu_B}{T} + \frac{\mu_C}{T} - \frac{\mu_D}{T}\right)$$

By writing this entropy production as a sum of a scalar and a vectorial part, $\sigma = \sigma_{vect} + \sigma_{scal}$, we find

$$\sigma_{vect} = -J_B\frac{\partial}{\partial x}\left(\frac{\mu_B}{T}\right) - J_C\frac{\partial}{\partial x}\left(\frac{\mu_C}{T}\right) - J_D\frac{\partial}{\partial x}\left(\frac{\mu_D}{T}\right)$$

$$\sigma_{scal} = \left(\frac{\mu_B}{T} + \frac{\mu_C}{T} - \frac{\mu_D}{T}\right)r = -\frac{r\Delta_r G}{T} = \sigma_{chem}$$

The vectorial contributions are due to diffusion while the scalar contribution is due to the reaction.

## 3.4   The frame of reference for fluxes

We need a frame of reference for the fluxes when we want to measure transport. In electroneutral systems, the electric current density, $j$, is constant throughout the system and independent of the frame of reference. Mass fluxes, on the other hand, depend on the velocity of the frame of reference. The solvent frame of reference is used when the movement of solutes with respect to the solvent is of interest. In flow problems, the center of mass frame of reference is convenient, see Chapter 6 and Appendix A.1. Transport across phase boundaries is technically important, and the frame of reference that gives a simple description is the surface itself [31]. In this frame of reference the observer moves along with the surface. If the surface is at rest, so is the observer. This is then also the laboratory frame of reference.

Some frames of references are defined below, to show how one description can be converted into another. In the conversion we take advantage of the fact that the entropy production is independent of the frame of reference. The notions, reversible and irreversible, are also independent of the frame of reference. They are, in other words, Galilei invariant. We may therefore convert all fluxes and conjugate forces from any frame of reference to another frame of reference and back without changing the entropy production, $\sigma$. According to Prigogine's theorem, any frame of reference that moves with a constant velocity with respect to the laboratory frame of reference, can be used for mass fluxes at mechanical equilibrium.

The mass flux of a component A relative to a frame of reference can be written as

$$J_{A,ref} = c_A(v_A - v_{ref}) \tag{3.17}$$

where $c_A$ is the concentration in mole/m$^3$, $v_A$ is the velocity of A and $v_{ref}$ is the velocity of the frame of reference relative to the laboratory frame of reference.

*The laboratory frame of reference or the wall frame of reference has:*

$$J_A = c_A v_A \quad \text{and} \quad v_{ref} = 0 \tag{3.18}$$

This is a convenient experimental frame of reference.

*The solvent frame of reference.* This frame of the reference is typically used when there is an excess of one component, the solvent. For transport of component A relative to the solvent one has:

$$J_{A,solv} = c_A(v_A - v_{solv}) \quad \text{and} \quad v_{ref} = v_{solv} \tag{3.19}$$

The frame of reference moves with the velocity of the solvent, $v_{solv}$.

*The average molar frame of reference.* In a multicomponent mixture, when there is no excess of one component, one can use the average molar velocity, which is defined by:

$$v_{molar} \equiv \frac{1}{c} \sum_i c_i v_i = \sum_i x_i v_i \qquad (3.20)$$

where $x_i = c_i/c$ is the mole fraction of $i$. This gives

$$J_{A,molar} \equiv c_A (v_A - v_{molar}) \qquad (3.21)$$

*The average volume frame of reference* has been used when transport occurs in a closed volume. The average volume velocity is

$$v_{vol} \equiv \sum_i c_i V_i v_i \qquad (3.22)$$

The flux of A becomes

$$J_{A,vol} \equiv c_A (v_A - v_{vol}) \qquad (3.23)$$

*The barycentric (average center of mass) frame of reference.* This frame of reference is used in the Navier-Stokes equation, Eq. (6.21). The average mass velocity is

$$v_{bar} \equiv \frac{1}{\rho} \sum_i \rho_i v_i \qquad (3.24)$$

where $\rho$ is the mass density of the fluid, and $\rho_i$ are the partial mass densities. The diffusion flux of A (in $mol.s^{-1}.m^{-2}$) in a barycentric frame of reference is

$$J_{A,bar} \equiv c_A (v_A - v_{bar}) \qquad (3.25)$$

The diffusion flux of A in $kg.s^{-1}.m^{-2}$ is equal to $M_A J_{A,bar} \equiv \rho_A (v_A - v_{bar})$, where $M_A$ is the molar mass. The advantage of the barycentric frame of reference for fluid dynamics is illustrated in Chapter 6.

# Chapter 4

# Flux equations and transport coefficients

*We present examples of flux equations that follow from the entropy production in Chapter 3. The determination of Onsager coefficients from experimental data, for instance, conductivities, diffusion coefficients and transport numbers, is discussed for simple homogeneous systems. We show that the coupling coefficients can be related to stored energy or work.*

In Chapter 3, we derived the entropy production for systems in mechanical equilibrium, using the assumption of local equilibrium. The entropy production determines the conjugate thermodynamic forces and fluxes of the system. The next major assumption in non-equilibrium thermodynamics is the assumption of linear flux-force relations. From this assumption, and the assumption of microscopic reversibility, Onsager derived the symmetry relation, Eq. (1.3), for the coupling coefficients. He also gave the fundamental relation between a flux and a force, and explained why they can be called conjugate [2, 3].

In this Chapter, we give examples of flux equations in order to illustrate the meaning of a characteristic property of the theory; the coupling coefficient. The transport problems we are dealing with, can be well studied in one dimension. Shear flow, which has more dimensions, is dealt with in Chapter 6. The general aspects discussed in Section 4.1, apply also to Chapter 6, however. Transport into and through surfaces is discussed in Chapters 9 and 10.

## 4.1   Linear flux-force relations

Irreversible thermodynamics assumes that all fluxes, $J_i$, are linear functions of all forces, $X_j$. In general:

$$J_i = \sum_{j=1}^{n} L_{ij} X_j \qquad \text{for} \qquad j = 1, 2, ..., n \qquad (4.1)$$

where $n$ is the number of *independent* fluxes. The coefficients must be determined from experiments. We assume that the linear relations are valid *locally*. All coefficients are functions of the local values of the state variables of the system. A global description of the system might thus give a nonlinear relation between $J_i$ and integrated (overall) forces. Locally, the conductivities $L_{ij}$'s *do not depend on the* $X_j$'s, however. The description is linear only in this sense. It means, for instance, that the thermal conductivity is allowed to depend on the temperature, but not on the gradient of the temperature.

**Remark 4** *The theory is often called "linear non-equilibrium thermodynamics". Contrary to what is sometimes stated, the theory results in an extremely nonlinear description of the system. Phenomena like turbulence and the Rayleigh-Benard instability, as described with the Navier-Stokes equation, cf. Eq. (6.21), occur within the framework of linear non-equilibrium thermodynamics. The use of the adjective "linear" is therefore rather misleading. See the 2nd preface of the Dover edition of de Groot and Mazur [12].*

Alternatively, the forces may be expressed as linear functions of all fluxes

$$X_k = \sum_{j=1}^{n} R_{kj} J_j \qquad k = 1, 2, ..., n \qquad (4.2)$$

The resistivity matrix is the inverse of the conductivity matrix

$$\sum_{i=1}^{n} L_{ik} R_{kj} = \sum_{i=1}^{n} R_{ik} L_{kj} = \delta_{ij} \qquad (4.3)$$

Here $\delta_{ij}$ is a Kronecker delta, which is one if the indices are equal, and zero otherwise. Equations (4.1) and (4.2) are equivalent. It depends on the system which formulation is more convenient to use.

Onsager [2, 3] proved that the matrix of coefficients for Eq. (4.1) is symmetric when the system is microscopically reversible.[1]

$$L_{ik} = L_{ki} \quad \text{or equivalently} \quad R_{ik} = R_{ki} \qquad (4.4)$$

We shall not repeat Onsager's derivation. Førland *et al.* [19] gave a simple presentation of the original work. The Onsager relations simplify the transport problem. When the *Onsager conductivity* coefficients $L_{ik}$ are known, we know how the different processes are coupled to one another. "Being coupled" here, is used solely in the sense that a force $X_k$ leads to a flux $J_i$, and vice versa. A temperature gradient can, for instance, give rise to a mass flux, and a chemical potential gradient can give rise to a heat flux.

Before we study examples, we examine some general properties of the Onsager coefficients. By substituting the linear laws (4.1) into the entropy production, we have

$$\sigma = \sum_i X_i \sum_k L_{ik} X_k = \sum_{i,k} X_i L_{ik} X_k \geq 0 \qquad (4.5)$$

Similarly, we find by introducing Eq. (4.2)

$$\sigma = \sum_{i,k} J_i R_{ik} J_k \geq 0 \qquad (4.6)$$

The inequalities follow from the second law of thermodynamics. By taking all the forces, except one, equal to zero, it follows immediately that main coefficients are always positive

$$L_{ii} \geq 0 \quad \text{and} \quad R_{ii} \geq 0 \qquad (4.7)$$

---

[1]Microscopic reversibility is the following property: The probability that a fluctuation $\alpha_i$ in some property at time $t$, is followed by a fluctuation $\alpha_j$ in another property after a time lag $\tau$ is equal to the probability of the reverse situation, where the fluctuation $\alpha_j$ at time $t$, is followed by $\alpha_i$ after a time lag $\tau$. This property is a consequence of the time reversal invariance of Newton's equations.

It follows for two pairs of fluxes and forces that:

$$L_{ii}L_{kk} - L_{ik}L_{ki} \geq 0 \quad \text{and} \quad R_{ii}R_{kk} - R_{ik}R_{ki} \geq 0 \qquad (4.8)$$

Transport is thus dominated by the main coefficients. The inequalities in Eq. (4.8) can be found by taking all except two forces as zero, and subsequently eliminating one of the two forces in terms of the corresponding flux. For the conductivities, we obtain

$$J_1 = L_{11}X_1 + L_{12}X_2$$
$$J_2 = L_{21}X_1 + L_{22}X_2 \qquad (4.9)$$

By expressing $X_2$ by the second equation and introducing it into the first, we obtain

$$J_1 = \left( L_{11} - \frac{L_{12}L_{21}}{L_{22}} \right) X_1 + \frac{L_{12}}{L_{22}} J_2 \qquad (4.10)$$

When this is introduced into the entropy production, we find:

$$\sigma = \left( L_{11} - \frac{L_{12}L_{21}}{L_{22}} \right) X_1^2 + \frac{L_{12}}{L_{22}} J_2 X_1 - \frac{L_{21}}{L_{22}} J_2 X_1 + \frac{J_2^2}{L_{22}}$$
$$= \left( L_{11} - \frac{L_{12}L_{21}}{L_{22}} \right) X_1^2 + \frac{J_2^2}{L_{22}} \qquad (4.11)$$

We must have $\sigma \geq 0$, also for the special case that $J_2 = 0$. Equation (4.8) for the conductivities follows from this and the last equality. The inequalities for the resistivities can be derived analogously. Two terms in the first equality cancel because of the Onsager relations. We shall see in Sections 4.2, 4.3 and 4.4 that the terms can be allocated to work performed and internal energy converted.

**Remark 5** *When $X_1 = 0$, a finite flux $J_2$ gives rise to a flux $J_1 = (L_{12}/L_{22})J_2$ in Eq. (4.10). The expression for the entropy production Eq. (4.11) shows, however, that a contribution to $J_1$ due to $J_2$ does not contribute to the entropy production. This transport along with $J_2$ is therefore reversible. The coupling coefficients describe reversible contributions to $\sigma$.*

An equality sign in Eq. (4.8) implies that two fluxes are proportional to one another. In the remark $J_1 = (L_{12}/L_{22})J_2$. A consequence is that the number of fluxes, and therefore the number of variables, can be reduced in the entropy production. By keeping the flux, $J_2$, rather than the force, $X_2$, equal to zero, the entropy production is reduced from $L_{11}X_1^2$ to $(L_{11} - L_{12}L_{21}/L_{22})X_1^2$. The combination of coefficients is an effective conductivity, see Exercise 2.2 in Section 2.2.

All cases studied in this Chapter can be cast into the general pattern above. We shall see how we can recognize the general pattern in the particular cases below.

## 4.2 Transport of heat and mass

Mass is transported, not only by a gradient in chemical potential. A gradient in temperature, can also lead to mass transport. This is called thermal diffusion or the Soret effect. Likewise, the gradient in chemical potential, leads to a heat flux; the so-called Dufour effect.

Consider a two-component system with heat and mass transfer. The second component has a concentration much larger than the first. It is then convenient to use the velocity of this component as the frame of reference for the velocity of the first component. This is the solvent frame of reference, cf. Section 3.4. Assume that the heat- and mass-transport take place in the $x$-direction only. The entropy production is, see Eq. (3.15),

$$\sigma = J_q' \frac{\partial}{\partial x}\left(\frac{1}{T}\right) + J_1\left(-\frac{1}{T}\frac{\partial \mu_{1,T}}{\partial x}\right) \tag{4.12}$$

The linear flux-force relations for the measurable heat flux and the mass flux are

$$J_q' = l_{qq}\frac{\partial}{\partial x}\left(\frac{1}{T}\right) + l_{q\mu}\left(-\frac{1}{T}\frac{\partial}{\partial x}\mu_{1,T}\right)$$

$$J_1 = l_{\mu q}\frac{\partial}{\partial x}\left(\frac{1}{T}\right) + l_{\mu\mu}\left(-\frac{1}{T}\frac{\partial}{\partial x}\mu_{1,T}\right) \tag{4.13}$$

where according to the Onsager relations $l_{q\mu} = l_{\mu q}$. Lower case $l$'s are used to describe diffusive transport of heat and mass. The chemical potential for a non-electrolyte is:

$$\mu_1 = \mu_1^\circ + RT \ln \gamma_1 c_1 \tag{4.14}$$

Table 4.1: Some interdiffusion coefficients at 300 K [36].

| System | $D_{1,2}$ m$^{-2}$s |
|---|---|
| CH$_4$ in nitrogen | $5 \times 10^{-5}$ |
| NaCl in water | $1.3 \times 10^{-9}$ |
| Sucrose in water | $4.5 \times 10^{-10}$ |
| C in steel | $1 \times 10^{-20}$ |

where $c_1$ is the concentration, $\gamma_1$ is the activity coefficient of component 1, and $\mu_1^{\circ}$ is the standard chemical potential, see Appendix A.3. When the temperature is constant, the equation for the mass flux (in mol.m$^{-2}$.s$^{-1}$) is

$$J_1 = -l_{\mu\mu} \frac{1}{T} \frac{\partial \mu_{1,T}}{\partial x}$$

The expression can then be related to Fick's law by

$$J_1 = -l_{\mu\mu} \frac{1}{T} \frac{\partial \mu_{1,T}}{\partial x} = -l_{\mu\mu} \frac{1}{T} \frac{\partial \mu_{1,T}}{\partial c_1} \frac{\partial c_1}{\partial x} = -D_{1,2} \frac{\partial c_1}{\partial x} \qquad (4.15)$$

where the *interdiffusion coefficient* of the first component $D_{1,2}$ can be identified with

$$D_{1,2} = \frac{l_{\mu\mu}}{T} \left[ \frac{\partial \mu_{1,T}}{\partial c_1} \right] = \frac{l_{\mu\mu} R}{c_1} \left[ 1 + \frac{\partial \ln \gamma_1}{\partial \ln c_1} \right] \qquad (4.16)$$

For ideal mixtures the parenthesis is unity. Interdiffusion coefficients vary by orders of magnitude when one compares gas, liquid, or solid phases, see Table 4.1. Even within the same phase, Fick's diffusion coefficient can vary by orders of magnitude. The variation of the Maxwell-Stefan diffusion coefficients (see Chapter 5) is less pronounced. Diffusion is often a rate-limiting process, not only in chemical reactors.

**Remark 6** *Interdiffusion coefficients are measured by spectroscopic and analytical techniques. When the coefficient is not known, one can obtain an estimate from the self-diffusion coefficient, $D_s$, which is measured using NMR. The self-diffusion coefficient is also given for pure components. It describes the Brownian motion of a single particle. Einstein gave the self-diffusion coefficient as $D_s = <x^2>/2t$, in terms of the mean square displacement $<x^2>$ during the time t. The self-diffusion coefficient of liquid water (in water) is*

$2.3 \times 10^{-9} m^2.s^{-1}$ at 300 K. For gases, kinetic theory gives $D_s = \ell v/3$, where $\ell$ is the mean free path of the molecule and $v$ is the mean thermal velocity. With a typical value $\ell = 30$ nm, and $v = 300$ m.s$^{-1}$, $D_s = 3 \times 10^{-6}$ m$^2.s^{-1}$. These estimates can be compared to the interdiffusion coefficients in Table 4.1. For low density gas mixtures, the interdiffusion and the self-diffusion coefficient are the same.

**Exercise 4.2.1** *How do you think the interdiffusion coefficient varies with increasing concentration of component 1 in a mixture?*

- **Solution:** From the expression for the interdiffusion coefficient given above we know that $D_{1,2} = l_{\mu\mu} \partial \mu_{1,T}/T \partial c_1$. For low densities, this reduces to $D_{1,2} = l_{\mu\mu} R/c_1$. For higher densities this becomes

$$D_{1,2} = \frac{l_{\mu\mu} R}{c_1} \left[ 1 + \partial \ln \gamma_1 / \partial \ln c_1 \right]$$

  The concentration dependence of $l_{\mu\mu}$ must be measured.

When the chemical potential at constant temperature, $\mu_{1,T}$, is constant, the equation for the heat flux reduces to Fourier's law for a homogeneous system

$$J_q' = l_{qq} \frac{\partial}{\partial x} \left( \frac{1}{T} \right) = -\lambda_\mu \frac{\partial T}{\partial x} \tag{4.17}$$

This gives the following relation between $l_{qq}$ and the thermal conductivity of a homogeneous material:

$$\lambda_\mu = - \left( \frac{J_q'}{\partial T/\partial x} \right)_{\partial \mu/\partial x = 0} = \frac{1}{T^2} l_{qq} \tag{4.18}$$

This is not the conductivity measured in a two-component system at stationary-state, see Eq. (4.27) below. In the stationary state, internal energy and mass densities are independent of time. As a consequence, their fluxes are not only independent of time, but also independent of position. In the present example the total heat flux, $J_q$, and the mass flux, $J_1$, are constant both with respect to time and position. The measurable heat flux $J_q' = J_q - H_1 J_1$ is not necessarily constant as a function of position, however, due to the variation of the enthalpy density with local temperature and molar densities.

The Soret effect is mass transport that takes place due to $\partial T/\partial x$. For a constant chemical potential the mass flux in the solvent frame of reference is given by Eq. (4.13) or

$$J_1 = l_{\mu q}\frac{\partial}{\partial x}\left(\frac{1}{T}\right) = -c_1 D_T \frac{\partial T}{\partial x} \qquad (4.19)$$

where the *thermal diffusion coefficient*, $D_T$, is defined by[2]

$$D_T = \frac{l_{\mu q}}{c_1 \left(T\right)^2} \qquad (4.20)$$

The ratio of the thermal diffusion coefficient and the interdiffusion coefficient is called the Soret coefficient, $s_T$. For the system in a stationary state such that $J_1 = 0$, the Soret coefficient can be expressed as the ratio of the concentration and the temperature gradients:

$$s_T \equiv -\left(\frac{\partial c_1/\partial x}{c_1 \partial T/\partial x}\right)_{J=0} = \frac{D_T}{D_{1,2}} \qquad (4.21)$$

By measuring the gradients in temperature and concentration, and the interdiffusion coefficient, one can calculate the thermal diffusion coefficient.

Heat transport due to a concentration gradient is called the *Dufour effect*. This effect is expressed by $l_{q\mu}$, and is the *reciprocal* of the Soret effect. The coefficient ratio

$$q^* = \left(\frac{J'_q}{J_1}\right)_{dT=0} = \frac{l_{q\mu}}{l_{\mu\mu}} \qquad (4.22)$$

is the *heat of transfer*. By using Eqs. (4.20) and (4.16), we can express the heat of transfer by the ratio of the thermal diffusion coefficient and the interdiffusion coefficient:

$$q^* = \frac{c_1 D_T T}{D_{1,2}}\left(\frac{\partial \mu_{1,T}}{\partial c_1}\right)$$

$$= s_T c_1 T\left(\frac{\partial \mu_{1,T}}{\partial c_1}\right) = s_T R T^2 \left[1 + \frac{\partial \ln\gamma_1}{\partial \ln c_1}\right] \qquad (4.23)$$

---

[2]Kuiken [25] used a different definition of the thermal diffusion coefficient, which is more appropriate in a multicomponent system where none of the components is present in excess.

Table 4.2: Soret coefficients at 300 K for some binary mixtures. The mole fraction, $x_1 = c_1/c$, refers to the first-mentioned component.

| System | $x_1$ | $T$ K$^{-1}$ | $p$ bar$^{-1}$ | $s_T$ K |
|---|---|---|---|---|
| Methane in propane | 0.34 | 346 | 60800 | 0.042 |
| Methane in cyclopentane | .0026 | 293 | 1 | -0.016 |
| i-butane in methylcyclopentane | 0.5 | 293 | 1 | -0.0096 |
| Cyclohexane in benzene | 0.5 | 293 | 1 | -0.0063 |
| Carbon dioxide in hydrogen | 0.51 | 223 | 15 | .00046 |

An experiment that gives the Soret coefficient, $s_T$, also gives the heat of transfer, $q^*$.

The Soret effect is known to damage materials that have their strength defined by small additives. For instance: the depletion of chromium, molybdenum as well as of carbon will weaken steel in nuclear reactors. A familiar example of the Soret effect can be observed at hot radiators in houses. There is convection, but also thermal diffusion. We see dust particles accumulate near the cold window, but not at the hot radiator. The Soret coefficient is normally positive for the light component and negative for the heavy component in a mixture. This explains that heavy components accumulate on cold sides, and light ones on the warm side. Holt *et al.* [52] studied the distribution of methane and decane in the geothermal gradient in an oil reservoir using this. Some values are given in Table 4.2 [53]. Fröba *et al.* [54] have reviewed measurement techniques.

The Soret coefficient in gases is usually small. Nevertheless, it has been used to separate isotopes, a difficult separation. Isotopes differ by their masses only ($m_2$ and $m_1$). Furry *et al.* [55] used the formula

$$s_T = \frac{0.35(m_2 - m_1)}{T(m_2 + m_1)}$$

to design separation columns for radioactive isotopes. This effect is not sufficient to explain chemical effects [56]. Kempers formula [57] for the thermal diffusion factor is now well-established for hydro-carbons:

$$\alpha_T \equiv s_T T = \frac{V_1 V_2}{x_1 V_1 + x_2 V_2} \frac{\frac{H_2 - H_2^0}{V_2} - \frac{H_1 - H_1^0}{V_1}}{x_1 \left( \partial \mu_1 / \partial x_1 \right)_{p,T}} + \frac{RT \alpha^0}{x_1 \left( \partial \mu_1 / \partial x_1 \right)_{p,T}}$$

(4.24)

The first identity defines the thermal diffusion factor. The value $\alpha^0$ can be calculated from kinetic theory. The partial molar enthalpies of the mixture ($H_i$) and of pure components ($H_i^0$), the partial molar volumes and the composition, and the thermodynamic factor (the derivative of the chemical potential with respect to composition), can all be determined from an equation of state of the fluid.

The heat transport due to mass transport is:

$$J_q' = -\frac{1}{T^2} \left( l_{qq} - \frac{l_{q\mu} l_{\mu q}}{l_{\mu\mu}} \right) \frac{\partial T}{\partial x} + q^* J$$

(4.25)

The part of the heat flux that contains $q^*$, changes direction with $J$, compare Eq. (4.10). The heat transport due to the Dufour effect is reversible in this sense. It leads to separation and therefore work. This property becomes more evident in thermal osmosis, *cf.* Chapter 10, where solutions are separated by a membrane.

Fourier's law

$$J_q' = -\lambda \frac{\partial T}{\partial x}$$

(4.26)

can also be used to identify the effective thermal conductivity for zero mass flux, $\lambda$. We obtain the heat flux for these conditions from Eq. (4.25) as

$$\lambda = - \left[ \frac{J_q'}{(\partial T / \partial x)} \right]_{J=0} = \frac{1}{T^2} \left( l_{qq} - \frac{l_{q\mu} l_{\mu q}}{l_{\mu\mu}} \right)$$

(4.27)

The thermal conductivity $\lambda$ can be found from plots like that shown in Fig. 4.1. It is generally found that $\lambda$ rather than $(l_{qq} - l_{q\mu} l_{\mu q} / l_{\mu\mu})$ is constant over a wide range of temperatures. By comparing Eq. (4.18) to Eq. (4.27), we find that $\lambda$ is smaller than $\lambda_\mu$. A temperature difference across a homogeneous system, leads initially to a heat flux and thus a particle flux. The entropy production is initially proportional to $\lambda_\mu$, when $\partial T / \partial x$ is constant and $\partial \mu_{1,T} / \partial x = 0$. The

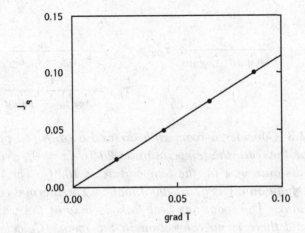

Figure 4.1: The dimensionless stationary state heat flux as a function of the dimensionless temperature gradient of a binary mixture obtained by molecular dynamics simulations [51]. The temperature gradient is of the order $10^8$ K.m$^{-1}$.

particle flux changes the chemical potential gradient until, in the resulting stationary state, the particle flux becomes zero. The entropy production becomes then proportional to $\lambda$. The entropy production in the stationary state is therefore smaller than in the initial state, *cf.* Eq. (4.11).

From the positive nature of the entropy production, it follows that the diffusion coefficient and the thermal conductivity are positive, *cf.* Eqs. (4.7)), (4.16) and (4.27). The thermal diffusion coefficient and the heat of transfer can be positive or negative. From Eq. (4.8) we have an upper bound on the absolute value of the thermal diffusion coefficient.

$$(D_T)^2 \leqslant D_1 \lambda_\mu \left( (c_1)^2 T \frac{\partial \mu_{1,T}}{\partial c_1} \right)^{-1} \tag{4.28}$$

**Exercise 4.2.2** *Express* $-\frac{1}{T^2} (\partial T/\partial x)$ *and* $-\frac{1}{T} \partial \mu_{1,T}/\partial x$ *in terms of* $J'_q$ *and* $J$.

- **Solution:** By inverting Eq. (4.13) we have:

$$-\frac{1}{T^2} \frac{\partial T}{\partial x} = r_{qq} J'_q + r_{q\mu} J \quad \text{and} \quad -\frac{1}{T} \frac{\partial \mu_{1,T}}{\partial x} = r_{\mu q} J'_q + r_{\mu\mu} J$$

with

$$r_{qq} = \frac{l_{\mu\mu}}{l_{qq}l_{\mu\mu} - l_{\mu q}l_{q\mu}}, \quad r_{\mu q} = r_{q\mu} = \frac{-l_{q\mu}}{l_{qq}l_{\mu\mu} - l_{\mu q}l_{q\mu}},$$

$$r_{\mu\mu} = \frac{l_{qq}}{l_{qq}l_{\mu\mu} - l_{\mu q}l_{q\mu}}$$

**Exercise 4.2.3** *Consider a room with air and a trace (10 ppm) of perfume. Near the wall, the temperature is $20\,^{\circ}C$ ($x = 0$). Near the window at a distance of 4 m, the temperature is $10\,^{\circ}C$. The heat of transfer of perfume in air is $q^* = 700$ J.mol.$^{-1}$ The thermal conductivity is constant. The room has heat leakage only at the wall with the window and there is no convection in the room. Calculate the concentration difference of the perfume between the window and the wall in the stationary state.*

- **Solution:** In the stationary state there is no perfume flux. This implies that, *cf.* Eqs. (4.25)) and (4.27)), $J_q' = -\lambda dT/dx$. Furthermore it follows from Eqs. (4.13)) and (4.22) that $d\mu_T/dx = -(q^*/T)dT/dx$. Because of the absence of perfume flux, the heat current is constant so that $T(x) = 20-2.5x$. This results in $d\mu_T/dx = (RT/c)dc/dx = 2.5q^*/T$, where the temperature is in K. The concentration gradient $dc/dx = cq^*/RT^2(dT/dx) \simeq 0.025$. This gives a concentration of 0.1 ppm higher at the window than close to the wall. Perfume concentrates at the cold side, it is heavier than air.

## 4.3   Transport of heat and charge

Coupled transports of heat and charge take place in semiconducting devices. To illustrate the coupling between fluxes of heat and charge, consider a piece of lead, which is connected to a potentiometer via molybdenum wires

$$\text{Mo}(T) \mid \text{Pb} \mid \text{Mo}(T + \Delta T) \qquad (4.29)$$

The joints are kept at different temperatures. The temperature changes continuously in the homogeneous phases and jumps at the surfaces. The electric current is defined as positive when positive charges are moving from left to right in the system. Such a system can be used to generate electric power from a waste heat source (the Seebeck effect). It can also be used for cooling purposes (the Peltier effect).

The entropy production for the homogeneous phases is obtained from Eq. (3.15). In this case, the equation reduces to

$$\sigma = -J_q' \frac{1}{T^2} \frac{\partial T}{\partial x} - j \frac{1}{T} \frac{\partial \phi}{\partial x} \qquad (4.30)$$

The flux equations are

$$J_q' = -L_{qq} \frac{1}{T^2} \frac{\partial T}{\partial x} - L_{q\phi} \frac{1}{T} \frac{\partial \phi}{\partial x}$$

$$j = -L_{\phi q} \frac{1}{T^2} \frac{\partial T}{\partial x} - L_{\phi\phi} \frac{1}{T} \frac{\partial \phi}{\partial x} \qquad (4.31)$$

The first flux equation says that heat can also be transported by means of an electric field. According to the last equation we can use a temperature gradient to generate a potential difference and an electric current. In order to define the coefficients, consider first isothermal conditions. The electric current density is:

$$j = -L_{\phi\phi} \frac{1}{T} \frac{\partial \phi}{\partial x} \qquad (4.32)$$

By equating this to Ohm's law, $j = -\kappa \partial \phi / \partial x$, we identify the electric conductivity

$$\kappa = \frac{L_{\phi\phi}}{T} \qquad (4.33)$$

The *Peltier coefficient*, $\pi$, is defined as the heat transferred *reversibly* with the electric current (*cf.* Remark 5) at constant temperature:

$$\pi \equiv F \left( \frac{J_q'}{j} \right)_{dT=0} = F \frac{L_{q\phi}}{L_{\phi\phi}} \qquad (4.34)$$

The coupling between heat and charge transport gives rise to a potential gradient, when there is a temperature gradient. From Eq. (4.31), we obtain

$$\frac{\partial \phi}{\partial x} = -\frac{L_{\phi q}}{L_{\phi\phi}} \frac{1}{T} \frac{\partial T}{\partial x} - \frac{j}{\kappa} \qquad (4.35)$$

The ratio $(\mathrm{d}\phi/\mathrm{d}T)_{j=0}$ is called the Seebeck coefficient. The Peltier and Seebeck coefficients are related by the Onsager relations:

$$F\left(\frac{\mathrm{d}\phi}{\mathrm{d}T}\right)_{j=0} = -\frac{\pi}{T} \qquad (4.36)$$

The Peltier coefficient can be interpreted as entropy that is *transported* by the charge carrier. For the molybdenum and lead phases, we have

$$\pi_\mathrm{i} \equiv F\left(\frac{J_q^{\prime \mathrm{i}}}{j}\right)_{\mathrm{d}T=0} = TS_{\mathrm{e}^-,\mathrm{i}}^* \qquad (4.37)$$

where i stands for Mo or Pb, respectively, and $S_{\mathrm{e}^-,\mathrm{i}}^*$ is the transported entropy.

The transported entropy is positive when entropy is transported along with positive charges. In metals, charge is mostly carried by electrons. Electronic conductors have transported entropies between $-1$ and $-20$ J.K$^{-1}$.mol$^{-1}$ [58]. Some transition metals like Mo, Cr and W have negative transported entropies, while Pb has a positive value. The maximum value for Mo is $S_{\mathrm{e}^-,\mathrm{Mo}}^* = -17$ J.K$^{-1}$.mol$^{-1}$ at 900 K [58]. At the same temperature, $S_{\mathrm{e}^-,\mathrm{Pb}}^* = 5$ J.K$^{-1}$.mol$^{-1}$. The transported entropy in Eq. (4.37) can therefore be positive or negative. Transported entropies are kinetic, not thermodynamic properties.

**Remark 7** *While the thermodynamic entropy is an absolute quantity, the transported entropy of a charge carrier is not. It depends on a choice of a reference compound, since only the combination of transported entropies enter the expression for the electromotive force. The transported entropy of electrons in lead is the commonly used reference value. This means that the entropy flux for a material depends on this reference. The net heat effect at the surface is absolute, however.*

By eliminating the potential gradient in Eq. (4.31), we can write the heat flux on the form given by Eq. (4.10):

$$J_q' = -\lambda\frac{\partial T}{\partial x} + \frac{\pi}{F}j \qquad (4.38)$$

where

$$\lambda \equiv -\left(\frac{J_q'}{\partial T/\partial x}\right)_{j=0} = \frac{1}{T^2}\left(L_{qq} - \frac{L_{\phi q}L_{q\phi}}{L_{\phi\phi}}\right) \qquad (4.39)$$

The factor $\lambda$ is the thermal conductivity when the electric current is zero, compare Eq. (4.27). Equation (4.38) expresses that a heat flux may arise due to not only a temperature gradient but also due to an electric current. This effect has been used to construct thermoelectric cooling devices. Cooling occurs particularly at junctions, when the transported entropy changes more than in the homogeneous conductor. In the example above, there is a Peltier effect at 900 K of $(-17 - 5)$ J.K$^{-1}$.mol$^{-1}$.900 K $= -$ 19.8 kJ.mol$^{-1}$. In the presence of an electric current of 1 A.m$^{-2}$.s$^{-1}$, the cooling effect is 0.11 J.m$^{-2}$.s$^{-1}$.

The heat flux due to the electric current reverses direction by reversing the current. In this manner we can turn a cooling device into a heating device.

**Exercise 4.3.1** *Consider the example described in this subsection. When we let an electric current of $10^4$ A.m$^{-2}$ run through the lead, entropy will be transported. If the system is thermally insulated, a temperature gradient will build up. Calculate the stationary temperature difference over a distance of 2 m for an insulated system. Use a stationary thermal conductivity of 5 J.K$^{-1}$. m$^{-1}$.s$^{-1}$ and an average temperature of 300 K. The transported entropy is 5 J.K$^{-1}$. mol$^{-1}$.*

- **Solution:** Eq. (4.38) gives the heat flux in terms of the temperature gradient and the electric current

$$J_q' = -\lambda_{Pb}\frac{\partial T}{\partial x} + \frac{\pi_{Pb}}{F}j$$

In the stationary state, the heat flux is zero and it follows that

$$\frac{dT}{dx} = \frac{TS_{e^-,Pb}^* j}{F\lambda_{Pb}} = \frac{300 \text{ K}(5 \text{ J.K}^{-1}.\text{mol}^{-1})10^4 \text{ A.m}^{-2}}{96500 \text{ C.mol}^{-1}.5 \text{ W.K}^{-1}.\text{m}^{-1}} = 31 \text{ K.m}^{-1}$$

Over a distance of 2 m this gives a temperature increase of 62 K.

**Exercise 4.3.2** *The same piece of lead is electrically isolated, and has a heat flux of 100 W.m$^{-2}$. Calculate the maximum electrical potential gradient that arise.*

- **Solution:** The measurable heat flux in Eq. (4.38) can also be written as

$$J_q' = -\lambda_{Pb}\frac{\partial T}{\partial x} + \frac{TS_{e^-,Pb}^*}{F}j$$

At the maximum electric potential difference $j = 0$, so that

$$J_q' = -\lambda_{Pb}\frac{\partial T}{\partial x}$$

Together with Eq. (4.31) it follows that

$$\frac{\partial\phi}{\partial x} = -\frac{S_{e^-,Pb}^*}{F}\frac{\partial T}{\partial x} = \frac{S_{e^-,Pb}^*}{F\lambda_{Pb}}J_q'$$

This gives

$$\frac{\partial\phi}{\partial x} = \frac{5 \text{ J.K}^{-1}.\text{mol}^{-1}.100 \text{ W.m}^{-2}}{96500 \text{ C.mol}^{-1}.5 \text{ W.K}^{-1}.\text{m}^{-1}} = 10^{-3} \text{ V.m}^{-1}$$

The ends of the left (l) and right (r) molybdenum wires have the temperature of the potentiometer, $T^{l,o} = T^{r,o} = T^o$. We neglect the temperature dependence of the transported entropies and calculate the contributions to the electromotive force. The lead conductor gives

$$(\Delta_m\phi)_{j=0} = \frac{1}{F}S_{e^-,Pb}^* \left(T^{m,r} - T^{m,l}\right) \qquad (4.40)$$

where m denotes the homogeneous lead phase. The left and right molybdenum wires give

$$(\Delta_l\phi)_{j=0} = \frac{1}{F}S_{e^-,Mo}^* \left(T^{l,m} - T^o\right)$$

$$(\Delta_r\phi)_{j=0} = \frac{1}{F}S_{e^-,Mo}^* \left(T^o - T^{r,m}\right) \qquad (4.41)$$

With $T^{l,m} = T^{m,l}$ and $T^{m,r} = T^{r,m}$, the sum is:

$$(\Delta\phi)_{j=0} = \frac{1}{F}(S_{e^-,Pb}^* - S_{e^-,Mo}^*) \left(T^{m,r} - T^{m,l}\right) \qquad (4.42)$$

This sum is the electromotive force of the system. The Seebeck coefficient becomes:

$$\left(\frac{\Delta\phi}{T^{m,r} - T^{m,l}}\right)_{j=o} = \frac{1}{F}(S_{e^-,Pb}^* - S_{e^-,Mo}^*) \qquad (4.43)$$

The Seebeck coefficient is frequently used to determine transported entropies. With a typical value for the Seebeck coefficient of 20 $J.K^{-1}.mol^{-1}$, a temperature difference of 100 K will generate a potential difference of 20 mV. This is a small effect, but it can be enlarged by coupling single elements in series. In this manner one can make use of industrial waste heat.

**Exercise 4.3.3** *Production of silicon requires temperatures above 1800 °C. There are therefore high-temperature heat losses to the surroundings from the furnace and during casting. Consider silicon casting where the molten metal is (at least) 300 K above room temperature. The casting gives rise to fumes. A fan to ventilate the room requires 5.9 W at 0.1 A. A thermoelectric module of Be-Te with Seebeck coefficient $3.82 \times 10^3$ V.K$^{-1}$ is available. a) What is the electric potential obtainable from the module? b) The module electric resistance is 1.8 Ω. How many modules are needed to run a fan?*

- **Solution:** a) Using

$$\left( \frac{\Delta\phi}{T_h - T_c} \right)_{j=o} = 3.82 \times 10^3 \text{ V.K}^{-1}$$

we obtain

$$\Delta\phi_{j=0} = 3.82 \times 10^{-3}\Delta T = 3.82 \times 10^{-3} \text{ V.K}^{-1} \times 300K = 1.15 \text{ V}$$

b) When the fan is running, the power of one module becomes

$$\Delta\phi = \Delta\phi_{j=0} - R\,j = 1.15V - 1.8 \, \Omega \times 0.1 \text{ A} = 0.97 \text{ V}$$

The power of N modules should be 5.9 W, thus

$$Nj\Delta\phi = 5.9 \text{ W}$$

$$N = \frac{5.9 \text{ W}}{0.97 \text{ V} \times 01 \text{ A}} = 60.8$$

61 modules are needed.

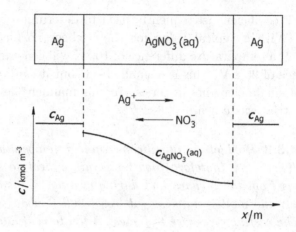

Figure 4.2: An isothermal electrochemical cell with a concentration gradient.

## 4.4   Transport of mass and charge

Coupled transports of mass and charge take place in all kinds of electrochemical cells, including in biological systems. Batteries and fuel cells generate electric potentials with order of magnitude 1 V, by invoking chemical reactions between the components of the cell. The coupled transports of mass and charge in concentration cells are simpler. In such cells, there is no spontaneous chemical reaction, and the electric potential is generated by concentration changes in the electrolyte. The resulting electric potential differences are small, but they are nevertheless interesting for exploitation, see Chapter 10. To illustrate the principles, take the isothermal *concentration cell*:

$$Ag(s)|AgNO_3(c_1)||AgNO_3(c_2)|Ag(s)$$

The electrodes are made of pure silver. The electrolyte is $AgNO_3$ in water, and there is a varying concentration of salt across the cell, see Fig. 4.2.

Consider the electrolyte. The entropy production Eq. (3.15) is

$$\sigma = J_{AgNO_3}\left(-\frac{1}{T}\frac{\partial \mu_{AgNO_3}}{\partial x}\right) + j\left(-\frac{1}{T}\frac{\partial \phi}{\partial x}\right) \qquad (4.44)$$

The temperature is constant throughout the system, and we have dropped subscript $T$ in $\mu_{AgNO_3}$. The flux equations are:

$$J_{AgNO_3} = -L_{\mu\mu}\frac{1}{T}\frac{\partial}{\partial x}\mu_{AgNO_3} - L_{\mu\phi}\frac{1}{T}\frac{\partial\phi}{\partial x}$$

$$j = -L_{\phi\mu}\frac{1}{T}\frac{\partial}{\partial x}\mu_{AgNO_3} - L_{\phi\phi}\frac{1}{T}\frac{\partial\phi}{\partial x} \qquad (4.45)$$

The frame of reference for the mass flux is the electrode surface. In this frame of reference, water is at rest. The solution is electroneutral, so that

$$c_{AgNO_3} = c_{Ag^+} = c_{NO_3^-} \qquad (4.46)$$

The chemical potential of $AgNO_3$ is

$$\mu_{AgNO_3} = \mu_{Ag^+} + \mu_{NO_3^-} = \mu_{AgNO_3}^0 + 2RT\ln\left(c_{AgNO_3}\gamma_\pm\right) \qquad (4.47)$$

where $\mu_{Ag^+}$ and $\mu_{NO_3^-}$ are the chemical potentials of the ions, $\gamma_\pm$ is the mean activity coefficient of the ions, equal for both ions, and $\mu_{AgNO_3}^0$ is the chemical potential of the salt in the standard state, see Appendix A.2. The electric conductivity for a uniform distribution of salt is

$$\kappa \equiv \frac{L_{\phi\phi}}{T} \qquad (4.48)$$

The coefficient $L_{\mu\mu}$ describes transport of $AgNO_3$ when the cell is short-circuited ($\Delta\phi = 0$). By eliminating the electric potential gradient with the help of Eq. (4.45), we obtain the salt flux

$$J_{AgNO_3} = -\frac{1}{T}\left(L_{\mu\mu} - \frac{L_{\phi\mu}L_{\mu\phi}}{L_{\phi\phi}}\right)\frac{\partial}{\partial x}\mu_{AgNO_3} + \frac{L_{\mu\phi}}{L_{\phi\phi}}j \qquad (4.49)$$

The diffusion coefficient from Fick's law, for zero electric current, is given by:

$$D_{AgNO_3} \equiv -\left(\frac{J_{AgNO_3}}{\partial c_{AgNO_3}/\partial x}\right)_{j=0} = \frac{1}{T}\left(L_{\mu\mu} - \frac{L_{\phi\mu}L_{\mu\phi}}{L_{\phi\phi}}\right)\frac{\partial\mu_{AgNO_3}}{\partial c_{AgNO_3}}$$

$$= \left(L_{\mu\mu} - \frac{L_{\phi\mu}L_{\mu\phi}}{L_{\phi\phi}}\right)\frac{2R}{c_{AgNO_3}}\left[1 + \frac{\partial\ln\gamma_\pm}{\partial\ln c_{AgNO_3}}\right]^{-1} \qquad (4.50)$$

The *transference coefficient* is defined as the ratio of the flux of salt and the electric current density at uniform composition:

$$t_{AgNO_3} \equiv F\left(\frac{J_{AgNO_3}}{j}\right)_{d\mu_{AgNO_3}=0} = F\frac{L_{\mu\phi}}{L_{\phi\phi}} \qquad (4.51)$$

Using these coefficients, the salt flux of Eq. (4.49) can be written as

$$J_{AgNO_3} = -D_{AgNO_3}\frac{\partial}{\partial x}c_{AgNO_3} + \frac{t_{AgNO_3}}{F}j \qquad (4.52)$$

Equation (4.52) expresses that diffusion and charge transfer are superimposed on one another. We described determinations of diffusion coefficients in Section 4.2 and give a method to estimate $D$ of electrolytes in Section 4.4.1.

The transference coefficient can be found in a *Hittorf* experiment, see e.g. [48]. In this experiment, an electric current is passing a cell of a uniform composition. A differential amount of salt will accumulate on one of the sides, and can be taken for analysis to determine the number of moles transported per moles of electric charge that is passing. The anode produces one mole of silver, and the cathode consumes one mole, per Faraday electrons passing the outer circuit. These changes plus the transport in the electrolyte, lead to a change in composition on both sides. The total flux of silver nitrate is given by Eq. (4.52). The equation says that in order to find $t_{AgNO_3}$ one must correct for the first term; diffusion. This can be done by measuring the concentration for small values of $j$ for a decreasing period of time and extrapolating to a zero period of time. The transference coefficient can be related to the transport number of one of the ions. The transport number is defined as the fraction of the electric current carried by the ion:

$$t_{Ag^+} = F\left(\frac{J_{Ag^+}}{j}\right)_{d\mu_{AgNO_3}=0}$$

$$t_{NO_3^-} = -F\left(\frac{J_{NO_3^-}}{j}\right)_{d\mu_{AgNO_3}=0} \qquad (4.53)$$

The two ions move in opposite directions, but together they are responsible for the total electric current, so that:

$$t_{Ag^+} + t_{NO_3^-} = 1 \qquad (4.54)$$

The electrodes of the present system are both reversible to $Ag^+$. The $NO_3^-$-ions do not leave the electrolyte, so the flux of the $Ag^+$ ions is therefore equal to $j/F$ plus the salt flux, while the flux of $NO_3^-$- ions is equal to minus the salt flux. The flux of $AgNO_3$ can therefore be defined by the flux of $NO_3^-$ ions [31]. As a consequence,

$$t_{AgNO_3} = -t_{NO_3^-} = -1 + t_{Ag^+} \tag{4.55}$$

**Remark 8** *The transference coefficient for the salt in the electrolyte are determined by the ions for which the reactions at the electrode surfaces are reversible. They are therefore not a material property of the electrolyte alone.*

The contribution from the electrolyte to the cell *potential* is obtained by solving Eq. (4.45), using $L_{\phi\mu} = L_{\mu\phi}$

$$\Delta\phi = -\frac{L_{\phi\mu}}{L_{\phi\phi}} \int_1^2 d\mu_{AgNO_3} - \frac{jd}{\kappa} = \frac{2t_{NO_3^-}}{F} \int_1^2 RT d\ln c_{AgNO_3} - \frac{jd}{\kappa} \tag{4.56}$$

where $d$ is the electrolyte thickness. This is the cell potential of a concentration cell (a cell with identical electrodes). We see that the coupling coefficient provides electric work. The potential at $j = 0$ is the reversible chemical work done by the system on expense of a lowering of its internal energy, *cf.* Eq. (4.11). This can be seen by introducing Eq. (4.45) into Eq. (4.44) and comparing compensating terms.[3] This expression can also be used to find $t_{AgNO_3}$ when $j = 0$ and the chemical potential in the two electrode compartments are known. This gives a more accurate determination than that obtainable from the Hittorf experiment, since the electric potential can be measured with a higher accuracy than concentration changes. The work obtainable from concentration cells can be produced in salt power plants, see Chapter 10.

## Exercise 4.4.1

*You have the following concentration cell:*

$$Ag(s)|\ AgCl(s)|\ KCl(aq,c_1)\ \|\ KCl(aq,c_2)|\ AgCl(s)|\ Ag(s).$$

---

[3]The electromotive force of the concentration cell is related to reversible phenomena only, and the common name of this potential, the "diffusion potential", is thus misleading.

*calculate the emf of a cell with $c_1 = 0.1$ $kmol^{-3}$ and $c_2 = 0.01$ $kmol.m^{-3}$ at $25\,°C$ and transference numbers $t_{K^+} = t_{Cl^-} = 0.5$.*

- **Solution:** When $j = 0$, the flux equation for charge can be written as

$$\frac{\partial \phi}{\partial x} = -\frac{L_{\phi\mu}}{L_{\phi\phi}}\frac{\partial \mu}{\partial x} = -\frac{2RT t_{KCl}}{F}\frac{\partial}{\partial x}\ln c$$

The emf is therefore

$$E = -\frac{2t_{KCl}RT}{F}\ln\frac{c_2}{c_1}$$

With $t_{KCl} = t_{K^+} = 0.5$,

$$E = -\frac{2 \times 0.5 \times 8.31 \text{ J.K}^{-1}.\text{mol}^{-1} \times 298 \text{ K}}{96500 \text{ C.mol}^{-1}}\ln\frac{0.01}{0.1} = 0.059 \text{ V}$$

By adding a perfect cation exchange membrane between the electrodes, we double this value, see Chapter 10.

### 4.4.1　The mobility model

A common model for the Onsager coefficients of an electrolytic solution uses mobilities of ions. In the example above, the mobilities are $u_{Ag^+}$ and $u_{NO_3^-}$. The assumption now is that the chemical potential gradient is equally effective as a force for transport as the electric potential gradient (Nernst-Einstein's assumption). In terms of symbols used here, this gives $FL_{\mu\mu} = -L_{\mu\phi}$. The mobility of an ion is defined as the ratio between the stationary velocity of the ion in a constant electric field, divided by the electric field [48]. For the present example, we find

$$L_{\mu\mu} = \frac{T}{F}c_{NO_3^-}u_{NO_3^-}, \qquad L_{\phi\mu} = L_{\mu\phi} = -Tc_{NO_3^-}u_{NO_3^-}$$

$$L_{\phi\phi} = FT\left(c_{NO_3^-}u_{NO_3^-} + c_{Ag^+}u_{Ag^+}\right) \tag{4.57}$$

Together with Eq. (4.46), the electric conductivity becomes

$$\kappa = Fc_{AgNO_3}\left(u_{NO_3^-} + u_{Ag^+}\right) \tag{4.58}$$

Table 4.3: Transference numbers and mobilities at infinite dilution at 298 K [59].

| Salt, MX | $t_{M^+}$ | $(\kappa_{MX}/c_{MX})$ $10^{-2}$ohm.mol.m$^{-2}$ | $u_{M^+}$ $10^{-8}$m$^{-2}$.s.V |
|---|---|---|---|
| HCl | 0.8209 | 426.16 | 36.23 |
| LiCl | 0.3360 | 114.99 | 73.52 |
| NaCl | 0.3870 | 126.50 | 5.19 |
| KCl | 0.4905 | 149.85 | 7.62 |
| KBr | 0.4846 | 151.67 | not available |

and the transference coefficient for the salt is, *cf.* Eq. (4.51),

$$t_{AgNO_3} = -\frac{u_{NO_3^-}}{u_{NO_3^-} + u_{Ag^+}} \tag{4.59}$$

The diffusion coefficient for the salt becomes

$$D_{AgNO_3} = \frac{2RT u_{Ag^+} u_{NO_3^-}}{F(u_{Ag^+} + u_{NO_3^-})} \tag{4.60}$$

Experimental data show that the diffusion coefficient is weakly dependent on the concentration. This can be explained by the mobility ratio being weakly dependent on the concentration of the salt. Some numbers from Ref. [59] are given in Table 4.3. The above equations in terms of the ionic mobilities, even though derived for an ideal mixture, have a rather large range of validity.

## 4.5 Concluding remarks

In this Chapter we have studied the coupling between pairs of processes, in a total of three examples. We have seen that the entropy production becomes smaller in the presence of coupling, and that coupling coefficients are central for extraction of work. Real systems have more than one pair of coupled transport processes. In polymer electrolyte fuel cells, for instance, there is coupled transport of water, electric charge (protons) and heat. The same systematic procedure can still be used to describe these processes, see [60].

The available work, which was examined here for coupled transports of heat and mass, heat and charge, and mass and charge, is smaller than work obtained in combustion engines, batteries or fuel cells. Work generated in the absence of chemical reactions is generally smaller, *cf.* Chapter 10. It may nevertheless become important, in a systematic effort, to reduce excess entropy production beyond today's state-of-the art. Efforts should be made to find better systems than those presented here [61].

The efficiency of energy converting processes is central in engineering and in Chapter 11 we study how the excess entropy production can be reduced by changing the way some process units are operated. The coupling effects presented in this Chapter, could be used in combination with the procedure outlined in Chapter 11.

# Chapter 5

# Non-isothermal multi-component diffusion

*Maxwell-Stefan equations were derived before Onsager established non-equilibrium thermodynamics. The equations are important, because they place all components on the same footing, and contain diffusion coefficients which are relatively constant for homogeneous fluids. We show how the equations are compatible with non-equilibrium thermodynamics and give rules of transformations between alternatives frames of reference for transport.*

Diffusion takes place in all chemical mixtures. Diffusion is often rate-limiting for processes that occur in nature and in industry, and it is therefore of interest to have a good description of diffusion, in isothermal as well as in nonisothermal systems. The simplest equation that describes diffusion is Fick's law, *cf.* Eq. (2.2). Very often there is an effect on the diffusion of one component due to concentration gradients of other components. There is, in other words, coupling of the diffusion fluxes. Such coupling effects are most efficiently described using the Maxwell-Stefan equations. In this Chapter we will explain how the Maxwell-Stefan equations are obtained in the context of non-equilibrium thermodynamics and also how they are related to alternative descriptions of multi-component diffusion.

## 5.1   Isothermal diffusion

Consider an isothermal and isobaric three-component mixture. The entropy production is

$$
\begin{aligned}
\sigma &= J_A \left( -\frac{1}{T}\frac{\partial \mu_A}{\partial x} \right) + J_B \left( -\frac{1}{T}\frac{\partial \mu_B}{\partial x} \right) + J_C \left( -\frac{1}{T}\frac{\partial \mu_C}{\partial x} \right) \\
&= v_A \left( -\frac{c_A}{T}\frac{\partial \mu_A}{\partial x} \right) + v_B \left( -\frac{c_B}{T}\frac{\partial \mu_B}{\partial x} \right) + v_C \left( -\frac{c_C}{T}\frac{\partial \mu_C}{\partial x} \right) \quad (5.1)
\end{aligned}
$$

where the velocity of component $i$ is defined by $v_i \equiv J_i/c_i$. The phenomenological equations for the thermodynamic forces can now be written as

$$
\begin{aligned}
-\frac{c_A}{T}\frac{\partial \mu_A}{\partial x} &= r_{AA}c_A c_A v_A + r_{AB}c_A c_B v_B + r_{AC}c_A c_C v_C \\
-\frac{c_B}{T}\frac{\partial \mu_B}{\partial x} &= r_{BA}c_B c_A v_A + r_{BB}c_B c_B v_B + r_{BC}c_B c_C v_C \\
-\frac{c_C}{T}\frac{\partial \mu_C}{\partial x} &= r_{CA}c_C c_A v_A + r_{CB}c_C c_B v_B + r_{CC}c_C c_C v_C \quad (5.2)
\end{aligned}
$$

The Onsager relations apply

$$
r_{AB} = r_{BA}, \quad r_{AC} = r_{CA}, \quad r_{BC} = r_{CB} \quad (5.3)
$$

According to Gibbs-Duhem's equation, the sum of the terms on the left hand side of these equations is zero

$$
c_A\frac{\partial \mu_A}{\partial x} + c_B\frac{\partial \mu_B}{\partial x} + c_C\frac{\partial \mu_C}{\partial x} = 0 \quad (5.4)
$$

Only two of the forces are therefore independent. As this is true for arbitrary velocities of the components, it follows that the resistivities in the matrix are dependent, and satisfy

$$
\begin{aligned}
r_{AA}c_A c_A + r_{BA}c_B c_A + r_{CA}c_C c_A &= 0 \\
r_{AB}c_A c_B + r_{BB}c_B c_B + r_{CB}c_C c_B &= 0 \\
r_{AC}c_A c_C + r_{BC}c_B c_C + r_{CC}c_C c_C &= 0 \quad (5.5)
\end{aligned}
$$

Using the Onsager relations it also follows that

$$r_{AA}c_Ac_A + r_{AB}c_Ac_B + r_{AC}c_Ac_C = 0$$
$$r_{BA}c_Bc_A + r_{BB}c_Bc_B + r_{BC}c_Bc_C = 0$$
$$r_{CA}c_Cc_A + r_{CB}c_Cc_B + r_{CC}c_Cc_C = 0 \qquad (5.6)$$

Onsager relations apply here also when the forces are dependent [12]. The resistivity matrix has an eigenvalue equal to zero, and thus a zero determinant. It can therefore not be inverted into a conductivity matrix. It follows from the above relations that there are only three independent resistivities. Once these have been obtained from experiments, the others can be calculated using the above relations. Similarly, one of the equations in Eq. (5.2) is minus the sum of the other two, and can be disregarded.

For an elegant discussion of the more general case when the system is neither isothermal nor isobaric, and when the system contains an arbitrary number of components, we refer to the monograph by Kuiken [25] and to the review by Krishna and Wesselingh [36].

### 5.1.1 Prigogine's theorem applied

In the linear laws, Eq. (5.2), we used the laboratory frame of reference. Other choices for the frame of reference are often used. Possible choices of a reference velocity, $v_{ref}$, are the center of mass, the average volume, the average molar or the solvent velocity, see Section 3.4.

According to Prigogine's theorem we can use an arbitrary frame of reference for the transports when the system is in mechanical equilibrium, see Ref. [12]. This gives

$$-\frac{c_A}{T}\frac{\partial \mu_A}{\partial x} = r_{AA}c_Ac_A\left(v_A - v_{ref}\right) + r_{AB}c_Ac_B\left(v_B - v_{ref}\right)$$
$$+ r_{AC}c_Ac_C\left(v_C - v_{ref}\right)$$
$$-\frac{c_B}{T}\frac{\partial \mu_B}{\partial x} = r_{BA}c_Bc_A\left(v_A - v_{ref}\right) + r_{BB}c_Bc_B\left(v_B - v_{ref}\right)$$
$$+ r_{BC}c_Bc_C\left(v_C - v_{ref}\right)$$
$$-\frac{c_C}{T}\frac{\partial \mu_C}{\partial x} = r_{CA}c_Cc_A\left(v_A - v_{ref}\right) + r_{CB}c_Cc_B\left(v_B - v_{ref}\right)$$
$$+ r_{CC}c_Cc_C\left(v_C - v_{ref}\right) \qquad (5.7)$$

These expressions follow from Eq. (5.2) using Eq. (5.6). By introducing $J_{i,\text{ref}} = c_i(\mathbf{v}_i - \mathbf{v}_{\text{ref}})$, the above equations become

$$-\frac{1}{T}\frac{\partial \mu_A}{\partial x} = r_{AA}J_{A,\text{ref}} + r_{AB}J_{B,\text{ref}} + r_{AC}J_{C,\text{ref}}$$

$$-\frac{1}{T}\frac{\partial \mu_B}{\partial x} = r_{BA}J_{A,\text{ref}} + r_{BB}J_{B,\text{ref}} + r_{BC}J_{C,\text{ref}}$$

$$-\frac{1}{T}\frac{\partial \mu_C}{\partial x} = r_{CA}J_{A,\text{ref}} + r_{CB}J_{B,\text{ref}} + r_{CC}J_{C,\text{ref}} \tag{5.8}$$

The same frame of reference, $\mathbf{v}_{\text{ref}}$, has been used. We can measure three independent resistivities in this frame of reference, and calculate the others using Eqs. (5.5) and (5.6), and subsequently use them in all other frames of reference. The procedure is explained in Section 5.1.2 with the solvent frame of reference as an example.

## 5.1.2 Diffusion in the solvent frame of reference

In this subsection we show how we can use one set of transport coefficients, determined in the solvent frame of reference, to calculate an equivalent set with a different frame of reference. The solvent frame of reference is used when there is an excess of one component, say component C. The velocity of the frame of reference is then $\mathbf{v}_{\text{ref}} = \mathbf{v}_{\text{solv}} = \mathbf{v}_C$. The first two flux equations in Section 5.1.1 reduce to

$$-\frac{1}{T}\frac{\partial \mu_A}{\partial x} = r_{AA}c_A\,(\mathbf{v}_A - \mathbf{v}_C) + r_{AB}c_B\,(\mathbf{v}_B - \mathbf{v}_C)$$

$$= r_{AA}J_{A,\text{solv}} + r_{AB}J_{B,\text{solv}}$$

$$-\frac{1}{T}\frac{\partial \mu_B}{\partial x} = r_{BA}c_A\,(\mathbf{v}_A - \mathbf{v}_C) + r_{BB}c_B\,(\mathbf{v}_B - \mathbf{v}_C)$$

$$= r_{BA}J_{A,\text{solv}} + r_{BB}J_{B,\text{solv}} \tag{5.9}$$

The resistivity matrix has been reduced to a symmetric two by two matrix which can be inverted. The fluxes relative to the solvent velocity become

$$J_{A,\text{solv}} = -l_{AA}\frac{1}{T}\frac{\partial \mu_A}{\partial x} - l_{AB}\frac{1}{T}\frac{\partial \mu_B}{\partial x}$$

$$J_{B,\text{solv}} = -l_{BA}\frac{1}{T}\frac{\partial \mu_A}{\partial x} - l_{BB}\frac{1}{T}\frac{\partial \mu_B}{\partial x} \tag{5.10}$$

where

$$l_{AA} = \frac{r_{BB}}{r_{AA}r_{BB} - r_{BA}r_{AB}} \ , \quad l_{AB} = l_{BA} = \frac{-r_{AB}}{r_{AA}r_{BB} - r_{BA}r_{AB}} \ ,$$

$$l_{BB} = \frac{r_{AA}}{r_{AA}r_{BB} - r_{BA}r_{AB}} \tag{5.11}$$

and vice-versa

$$r_{AA} = \frac{l_{BB}}{l_{AA}l_{BB} - l_{BA}l_{AB}} \ , \quad r_{AB} = r_{BA} = \frac{-l_{AB}}{l_{AA}l_{BB} - l_{BA}l_{AB}} \ ,$$

$$r_{BB} = \frac{l_{AA}}{l_{AA}l_{BB} - l_{BA}l_{AB}} \tag{5.12}$$

The entropy production becomes:

$$\sigma = J_{A,solv}\left(-\frac{1}{T}\frac{\partial \mu_A}{\partial x}\right) + J_{B,solv}\left(-\frac{1}{T}\frac{\partial \mu_B}{\partial x}\right)$$

$$= l_{AA}\left(\frac{1}{T}\frac{\partial \mu_A}{\partial x}\right)^2 + 2l_{AB}\left(\frac{1}{T}\frac{\partial \mu_A}{\partial x}\right)\left(\frac{1}{T}\frac{\partial \mu_B}{\partial x}\right) + l_{BB}\left(\frac{1}{T}\frac{\partial \mu_B}{\partial x}\right)^2$$

$$= r_{AA}J_{A,solv}^2 + 2r_{AB}J_{A,solv}J_{B,solv} + r_{BB}J_{B,solv}^2 \tag{5.13}$$

The fluxes have also been expressed in terms of the concentration gradients:

$$J_{A,solv} = -D_{AA,solv}\frac{\partial c_A}{\partial x} - D_{AB,solv}\frac{\partial c_B}{\partial x}$$

$$J_{B,solv} = -D_{BA,solv}\frac{\partial c_A}{\partial x} - D_{BB,solv}\frac{\partial c_B}{\partial x} \tag{5.14}$$

where the Fick diffusion coefficients are given by

$$D_{AA,solv} = l_{AA}\frac{1}{T}\frac{\partial \mu_A}{\partial c_A} = l_{AA}\frac{R}{c_A}$$

$$D_{AB,solv} = l_{AB}\frac{1}{T}\frac{\partial \mu_B}{\partial c_B} = l_{AB}\frac{R}{c_A}$$

$$D_{BA,solv} = l_{BA}\frac{1}{T}\frac{\partial \mu_A}{\partial c_A} = l_{BA}\frac{R}{c_B}$$

$$D_{BB,solv} = l_{BB}\frac{1}{T}\frac{\partial \mu_B}{\partial c_B} = l_{BB}\frac{R}{c_B} \tag{5.15}$$

*The matrix of Fick diffusion coefficients in the solvent frame of reference is not symmetric.* The second identity applies to the case of ideal solutions.

Once three of the four diffusion coefficients have been measured, we can determine the three conductivities $l_{AA}$, $l_{AB} = l_{BA}$ and $l_{BB}$. By using Eq. (5.12), we can next obtain the three independent resistivities. From the relations between the resistivities Eqs. (5.5 and 5.6) we can next obtain a complete matrix of (dependent) resistivities in Eq. (5.2). The procedure can be repeated with any other frame of reference. Alternatively, we can use the coefficients of Eq. (5.2) to calculate coefficients that belong to any frame of reference.

To summarize, Eq. (5.2) gives a common ground to transformations between different frames of reference. But a matrix with independent coefficients is needed to define experiments and determine the transport coefficients which are independent.

### 5.1.3   Maxwell-Stefan equations

Flux equations for isothermal mass transport were already written by Maxwell and Stefan in the nineteenth century, before Onsager established the theory of non-equilibrium thermodynamics, see [25] and [36]. The Maxwell-Stefan equations for multi-component diffusion have become popular for two reasons. The formulation uses velocity differences, which makes it independent of the choice of frame of reference, and the transport coefficients defined by these equations are relatively well-behaved if the fluid is homogeneous.

The Maxwell-Stefan equations can be derived from non-equilibrium thermodynamics by eliminating the main resistivities, $r_{AA}$, $r_{BB}$ and $r_{CC}$, in Eq. (5.2), using the identities in Eq. (5.6) giving

$$\frac{c_A}{T}\frac{\partial \mu_A}{\partial x} = r_{AB}c_A c_B \left(v_A - v_B\right) + r_{AC}c_A c_C \left(v_A - v_C\right)$$

$$\frac{c_B}{T}\frac{\partial \mu_B}{\partial x} = r_{BA}c_B c_A \left(v_B - v_A\right) + r_{BC}c_B c_C \left(v_B - v_C\right)$$

$$\frac{c_C}{T}\frac{\partial \mu_C}{\partial x} = r_{CA}c_C c_A \left(v_C - v_A\right) + r_{CB}c_C c_B \left(v_C - v_B\right) \qquad (5.16)$$

These were the equations written by Maxwell and Stefan. Only velocity differences of the components appear, making the description independent of the frame of reference.

By using Gibbs-Duhem's equation and the Onsager symmetry relations, it follows that the third equation is minus the sum of the other two equations. The above set of equations is therefore equivalent to

$$\frac{c_A}{T}\frac{\partial \mu_A}{\partial x} = r_{AB}c_A c_B \left(v_A - v_B\right) + r_{AC}c_A c_C \left(v_A - v_C\right)$$
$$\frac{c_B}{T}\frac{\partial \mu_B}{\partial x} = r_{AB}c_B c_A \left(v_B - v_A\right) + r_{BC}c_B c_C \left(v_B - v_C\right) \qquad (5.17)$$

These equations express two independent forces in two independent velocity differences. The Maxwell-Stefan diffusion coefficients are given in terms of the resistances by[1]

$$\mathcal{D}_{AB} = -\frac{R}{cr_{AB}} \qquad \mathcal{D}_{AC} = -\frac{R}{cr_{AC}} \qquad \mathcal{D}_{BC} = -\frac{R}{cr_{BC}} \qquad (5.18)$$

By using these diffusion coefficients, Eq.(5.17) becomes

$$-\frac{1}{RT}\frac{\partial \mu_A}{\partial x} = \frac{x_B}{\mathcal{D}_{AB}}\left(v_A - v_B\right) + \frac{x_C}{\mathcal{D}_{AC}}\left(v_A - v_C\right)$$
$$-\frac{1}{RT}\frac{\partial \mu_B}{\partial x} = \frac{x_A}{\mathcal{D}_{AB}}\left(v_B - v_A\right) + \frac{x_C}{\mathcal{D}_{BC}}\left(v_B - v_C\right) \qquad (5.19)$$

where $x_i \equiv c_i/c$ is the mole fraction of component $i$. The chemical potential is defined by, see Section A.3,

$$\mu_i = \mu_i^\circ(T) + RT \ln \gamma_i x_i \qquad (5.20)$$

where $\mu_i^0(T)$ is the standard state value and $\gamma_i$ is the activity coefficient. This gives

$$-\frac{\partial x_A}{\partial x}\left(1 + \frac{\partial \ln \gamma_A}{\partial \ln x_A}\right) = \frac{x_B}{\mathcal{D}_{AB}}\left(v_A - v_B\right) + \frac{x_C}{\mathcal{D}_{AC}}\left(v_A - v_C\right)$$
$$-\frac{\partial x_B}{\partial x}\left(1 + \frac{\partial \ln \gamma_B}{\partial \ln x_B}\right) = \frac{x_A}{\mathcal{D}_{AB}}\left(v_B - v_A\right) + \frac{x_C}{\mathcal{D}_{BC}}\left(v_B - v_C\right) \qquad (5.21)$$

---

[1]We follow [36] in this definition rather than [25] who used the pressure $p$ instead of $cRT$.

The Maxwell-Stefan diffusion coefficients are often found to be rather independent of the concentrations [25, 36]. *The Maxwell-Stefan diffusion coefficients are symmetric.* This is an important difference from the Fick diffusion coefficients, which are not symmetric.

The Fick diffusion coefficients are commonly found using experiments. Molecular dynamics simulations give the Maxwell-Stefan diffusion coefficients and the so-called self-diffusion coefficients. Self-diffusion coefficients can be measured using NMR. As one can understand from Eq. (5.15) the Fick and the Maxwell-Stefan diffusion coefficients can be expressed into each other if one knows the matrix of thermodynamic factors, $\partial \mu_i / \partial c_j$. The calculation of this matrix is not so easy. The recent development of the so-called Small System Method has improved this situation, see the review [62]. Fick's diffusion coefficients, which agree well with experiments, were found combining these results [63, 64].

The value of $\mathcal{D}_{AB}$ for any composition of a binary mixture of A and B has been estimated using the empirical Vignes rule

$$\mathcal{D}_{AB} = \left[\mathcal{D}_{AB(x_A \to 1)}\right]^{x_A} \cdot \left[\mathcal{D}_{AB(x_B \to 1)}\right]^{x_B} \tag{5.22}$$

which expresses the coefficient in terms of the values at infinitely dilution. These can be obtained from simulations or empirical relations. Improved formulae for multi-component diffusion were proposed by Liu and co-workers [62]. Using the assumption that the velocity cross-correlations are small compared to the velocity self-correlations they derived a multi-component Darken equation

$$\mathcal{D}_{ij} = D_{i,\text{self}} D_{j,\text{self}} \sum_{k=1}^{n} \frac{x_k}{D_{k,\text{self}}} \tag{5.23}$$

where $D_{k,\text{self}}$ is the self-diffusion coefficient of component $k$ in the mixture. For a binary mixture, $n = 2$, this equation reduces to the well-known Darken equation. In order to calculate the self-diffusion coefficient in the mixture, Liu *et al.* [62] proposed the following equation

$$D_{i,\text{self}} = \sum_{k=1}^{n} \frac{x_k}{D_{i,\text{self}}^{x_k \to 1}} \tag{5.24}$$

where $D_{i,\text{self}}^{x_k \to 1}$ is the self-diffusion coefficient of component $i$ in a $i, k$ binary mixture at infinite dilution. We refer to the review [62] for a detailed discussion and explanations on how to compute the self-diffusion coefficients from molecular dynamics simulations.

With computed values of the Maxwell-Stefan diffusion coefficients, we can calculate the resistivities of Eq. (5.18), which gives

$$r_{AB} = r_{BA} = -\frac{R}{c\mathcal{D}_{AB}} \quad , \quad r_{AC} = r_{CA} = -\frac{R}{c\mathcal{D}_{AC}} \quad ,$$
$$r_{BC} = r_{CB} = -\frac{R}{c\mathcal{D}_{BC}} \tag{5.25}$$

Equation (5.6) gives the main resistivities next.

The Maxwell-Stefan equations are symmetric for the interchange of components, contrary to all choices relative to some frame of reference. Therefore, there are $n(n-1)/2$ Maxwell-Stefan diffusion coefficients, but $(n-1)^2$ coefficients of all other choices. The fact that the Maxwell-Stefan diffusion coefficients are rather independent of the concentrations gives a possibility to predict the concentration dependence in, for instance, the solvent frame of reference. By knowing the diffusion coefficients in the solvent frame of reference for low concentrations, we can, following the scheme explained above, calculate the Maxwell-Stefan diffusion coefficients for low concentrations. Assuming these values to be correct for all concentrations, we can next proceed to calculate back and obtain reasonable approximations for the diffusion coefficients in the solvent frame of reference for all concentrations.

### 5.1.4 Changing a frame of reference

While the volume or the solvent frame of reference are convenient in analyzing experiments, the barycentric frame of reference is needed to describe pipe flow, see Chapter 6. It is therefore important to be able to convert from one frame of reference to another.

Most of the reference velocities given in Section 3.4 are averages of the velocities of the components and can be written as

$$v_{\text{ref}} \equiv a_A v_A + a_B v_B + a_C v_C \quad \text{with} \quad a_A + a_B + a_C = 1 \tag{5.26}$$

Table 5.1:   Coefficients for transformations between frames of reference.

|  | $a_A$ | $a_B$ | $a_C$ |
|---|---|---|---|
| solvent | 0 | 0 | 1 |
| average molar | $x_A$ | $x_B$ | $x_C$ |
| average volume | $c_A V_A$ | $c_B V_B$ | $c_C V_C$ |
| barycentric | $\rho_A/\rho$ | $\rho_B/\rho$ | $\rho_C/\rho$ |

Exceptions are the laboratory or the wall frame of reference, where $v_{ref} = 0$, and the surface frame of reference, which uses the velocity of the surface as a reference velocity. It follows from Eq. (5.26) that the mass fluxes satisfy

$$\frac{a_A}{c_A} J_{A,ref} + \frac{a_B}{c_B} J_{B,ref} + \frac{a_C}{c_C} J_{C,ref} = 0 \qquad (5.27)$$

They are therefore dependent in these frames of reference. For the various frames of reference, the averaging coefficients are given in Table 5.1. For any choice of the reference velocity, it is sufficient to use only two of the equations in Eq. (5.8). By using also Eq. (5.27), these two equations can be written as

$$-\frac{1}{T}\frac{\partial \mu_A}{\partial x} = \left( r_{AA} - \frac{a_A}{a_C}\frac{c_C}{c_A} r_{AC} \right) J_{A,ref} + \left( r_{AB} - \frac{a_B}{a_C}\frac{c_C}{c_B} r_{AC} \right) J_{B,ref}$$

$$\equiv r_{AA,ref} J_{A,ref} + r_{AB,ref} J_{B,ref}$$

$$-\frac{1}{T}\frac{\partial \mu_B}{\partial x} = \left( r_{BA} - \frac{a_A}{a_C}\frac{c_C}{c_A} r_{BC} \right) J_{A,ref} + \left( r_{BB} - \frac{a_B}{a_C}\frac{c_C}{c_B} r_{BC} \right) J_{B,ref}$$

$$\equiv r_{BA,ref} J_{A,ref} + r_{BB,ref} J_{B,ref} \qquad (5.28)$$

By inverting this equation, we obtain

$$J_{A,ref} = -l_{AA,ref} \left( \frac{1}{T}\frac{\partial \mu_A}{\partial x} \right) - l_{AB,ref} \left( \frac{1}{T}\frac{\partial \mu_B}{\partial x} \right)$$

$$J_{B,ref} = -l_{BA,ref} \left( \frac{1}{T}\frac{\partial \mu_A}{\partial x} \right) - l_{BB,ref} \left( \frac{1}{T}\frac{\partial \mu_B}{\partial x} \right) \qquad (5.29)$$

where

$$l_{AA,ref} = \frac{r_{BB,ref}}{r_{AA,ref}r_{BB,ref} - r_{BA,ref}r_{AB,ref}}$$

$$l_{AB,ref} = \frac{-r_{BA,ref}}{r_{AA,ref}r_{BB,ref} - r_{BA,ref}r_{AB,ref}}$$

$$l_{BA,ref} = \frac{-r_{AB,ref}}{r_{AA,ref}r_{BB,ref} - r_{BA,ref}r_{AB,ref}}$$

$$l_{BB,ref} = \frac{r_{AA,ref}}{r_{AA,ref}r_{BB,ref} - r_{BA,ref}r_{AB,ref}} \tag{5.30}$$

and vice-versa

$$r_{AA,ref} = \frac{l_{BB,ref}}{l_{AA,ref}l_{BB,ref} - l_{BA,ref}l_{AB,ref}}$$

$$r_{AB,ref} = \frac{-l_{BA,ref}}{l_{AA,ref}l_{BB,ref} - l_{BA,ref}l_{AB,ref}}$$

$$r_{BA,ref} = \frac{-l_{AB,ref}}{l_{AA,ref}l_{BB,ref} - l_{BA,ref}l_{AB,ref}}$$

$$r_{BB,ref} = \frac{l_{AA,ref}}{l_{AA,ref}l_{BB,ref} - l_{BA,ref}l_{AB,ref}} \tag{5.31}$$

We want to cast the diffusion equations in terms of concentration gradients, because these are often measured in diffusion experiments. By introducing the diffusion coefficients:

$$D_{AA,ref} = l_{AA,ref}\frac{1}{T}\frac{\partial\mu_A}{\partial c_A}$$

$$D_{AB,ref} = l_{AB,ref}\frac{1}{T}\frac{\partial\mu_B}{\partial c_B}$$

$$D_{BA,ref} = l_{BA,ref}\frac{1}{T}\frac{\partial\mu_A}{\partial c_A}$$

$$D_{BB,ref} = l_{BB,ref}\frac{1}{T}\frac{\partial\mu_B}{\partial c_B} \tag{5.32}$$

we can write Eq. (5.29) in the usual form

$$J_{A,ref} = -D_{AA,ref}\frac{\partial c_A}{\partial x} - D_{AB,ref}\frac{\partial c_B}{\partial x}$$

$$J_{B,ref} = -D_{BA,ref}\frac{\partial c_A}{\partial x} - D_{BB,ref}\frac{\partial c_B}{\partial x} \tag{5.33}$$

*The conductivity, the diffusivity and the resistivity matrices are, however, now all asymmetric!* Given the experimental values of these diffusion coefficients, we can calculate the conductivities $l_{ij}$ using Eq. (5.32). The two by two matrix of resistivities follows using Eq. (5.31). In order to find the six dependent resistivities, we use again the identities in Eq. (5.6), and the definitions of the resistivities in Eq. (5.28). The result is:

$$r_{AA,ref}c_Ac_A + r_{AB,ref}c_Ac_B + \frac{1}{a_C}r_{AC}c_Ac_C = 0$$

$$r_{BA,ref}c_Bc_A + r_{BB,ref}c_Bc_B + \frac{1}{a_C}r_{BC}c_Bc_C = 0$$

$$r_{CA}c_Cc_A + r_{CB}c_Cc_B + r_{CC}c_Cc_C = 0 \qquad (5.34)$$

After calculating $r_{AC} = r_{CA}$, $r_{BC} = r_{CB}$ and $r_{CC}$ with these expressions we can find $r_{AA}$, $r_{AB} = r_{BA}$ and $r_{BB}$ using the definitions of the resistivities in the frame of reference.

### Exercise 5.1.1

*When we use the solvent frame of reference, Fick's law can be written $J_{A,solv} = -D_{AA,solv}\partial c_A/\partial x$. With the average volume frame of reference, we write $J_{A,vol} = -D_{AA,vol}\partial c_A/\partial x$. For what condition is $D_{AA,vol} \simeq D_{AA,solv}$?*

- **Solution:** When $D_{AA,vol} \simeq D_{AA,solv}$ it follows using Fick's law that $J_{A,solv} \simeq J_{A,vol}$. For the solvent and average volume frames of reference we write

$$J_{A,solv} = c_A \left(v_A - v_{solv}\right)$$
$$J_{A,vol} = c_A \left(v_A - v_{vol}\right)$$

The fluxes and the diffusion constants are therefore approximately the same when

$$v_{solv} \simeq v_{vol}$$

Consider a two-component mixture where component 2 is the solvent. We then have

$$v_{solv} = v_2$$
$$v_{vol} = c_1V_1v_1 + c_2V_2v_2 = v_2 + c_1V_1\left(v_1 - v_2\right)$$

We see that $v_{solv} \simeq v_{vol}$, when $c_1 V_1 \ll c_2 V_2 \simeq 1$. This is often true for a dilute solution of component 1 in component 2.

## 5.2 Non-isothermal diffusion

It is interesting to generalize the Maxwell-Stefan form of the transport equations (*cf.* Section 5.1.3) to include also heat transport. Consider therefore again the coupled transport of three components in a system without external forces, but now in the presence of a temperature gradient.

The entropy production is according to Eq. (3.15)

$$\sigma = J'_q \left( -\frac{1}{T^2} \frac{\partial T}{\partial x} \right) + \sum_j^n J_i \left( -\frac{1}{T} \frac{\partial \mu_{j,T}}{\partial x} \right) \tag{5.35}$$

The flux equations in the resistivity-form become

$$-\frac{c_A}{T} \frac{\partial \mu_{A,T}}{\partial x} = r_{AA} c_A c_A v_A + r_{AB} c_A c_B v_B + r_{AC} c_A c_C v_C + r_{Aq} c_A J'_q$$

$$-\frac{c_B}{T} \frac{\partial \mu_{B,T}}{\partial x} = r_{BA} c_B c_A v_A + r_{BB} c_B c_B v_B + r_{BC} c_B c_C v_C + r_{Bq} c_B J'_q$$

$$-\frac{c_C}{T} \frac{\partial \mu_{C,T}}{\partial x} = r_{CA} c_C c_A v_A + r_{CB} c_C c_B v_B + r_{CC} c_C c_C v_C + r_{Cq} c_C J'_q$$

$$-\frac{1}{T^2} \frac{\partial T}{\partial x} = r_{qA} c_A v_A + r_{qB} c_B v_B + r_{qC} c_C v_C + r_{qq} J'_q \tag{5.36}$$

The Gibbs-Duhem equation

$$0 = -v dp + S dT + \sum_i x_i d\mu_i \tag{5.37}$$

plays a central role in the derivation of the Maxwell-Stefan equations. We have constant pressure, giving

$$0 = \sum_i \frac{c_i}{T} \frac{\partial \mu_{i,T}}{\partial x} \tag{5.38}$$

Equation (5.38) does not assume that the temperature is constant in the system. The differential chemical potential $d\mu_{i,T} = d\mu_i + (d\mu_i/dT)_{p,x_j} dT = d\mu_i - S_i dT$ is the change of chemical potential in the direction of constant temperature. The term $-S_i dT$ accounts for the variation in temperature; and this term, after summation over all components, cancels the term $SdT$ of Eq. (5.37), compare also Appendix A.2.

Using the Gibbs-Duhem equation, Eq. (5.38), we find that the sum of the first three equations of Eq. (5.36) is zero. Since the component velocities $v_A$, $v_B$, $v_C$, and the measurable heat flux $J'_q$ are not generally zero, the coefficients (resistivities) have to be correlated, and we obtain

$$r_{AA}c_Ac_A + r_{BA}c_Bc_A + r_{CA}c_Cc_A = 0$$
$$r_{AB}c_Ac_B + r_{BB}c_Bc_B + r_{CB}c_Cc_B = 0$$
$$r_{AC}c_Ac_C + r_{BC}c_Bc_C + r_{CC}c_Cc_C = 0$$
$$r_{Aq}c_A + r_{Bq}c_B + r_{Cq}c_C = 0 \qquad (5.39)$$

By using these relations and Gibbs-Duhem's equation, it follows that the third equation in Eq. (5.36) is minus the sum of the first two equations. The four equations are therefore equivalent to

$$-\frac{c_A}{T}\frac{\partial \mu_{A,T}}{\partial x} = r_{AA}c_Ac_Av_A + r_{AB}c_Ac_Bv_B + r_{AC}c_Ac_Cv_C + r_{Aq}c_A J'_q$$
$$-\frac{c_B}{T}\frac{\partial \mu_{B,T}}{\partial x} = r_{BA}c_Bc_Av_A + r_{BB}c_Bc_Bv_B + r_{BC}c_Bc_Cv_C + r_{Bq}c_B J'_q$$
$$-\frac{1}{T^2}\frac{\partial T}{\partial x} = r_{qA}c_Av_A + r_{qB}c_Bv_B + r_{qC}c_Cv_C + r_{qq}J'_q \qquad (5.40)$$

With the relations of Eq. (5.39) we can now eliminate the symmetric resistivities $r_{AA}c_Ac_A$, $r_{BB}c_Bc_B$, and $r_{qA}c_A$ from Eq. (5.40), and obtain

$$\frac{c_A}{T}\frac{\partial \mu_{A,T}}{\partial x} = r_{AB}c_Ac_B(v_A - v_B) + r_{AC}c_Ac_C(v_A - v_C) - r_{Aq}c_A J'_q$$
$$\frac{c_B}{T}\frac{\partial \mu_{B,T}}{\partial x} = r_{AB}c_Bc_A(v_B - v_A) + r_{BC}c_Bc_C(v_B - v_C) - r_{Bq}c_B J'_q$$
$$\frac{1}{T^2}\frac{\partial T}{\partial x} = r_{Bq}c_B(v_A - v_B) + r_{Cq}c_C(v_A - v_C) - r_{qq}J'_q \qquad (5.41)$$

where we have made use of the Onsager relations, $r_{AB} = r_{BA}$, $r_{AC} = r_{CA}$, $r_{BC} = r_{CB}$, $r_{qA} = r_{Aq}$, $r_{qB} = r_{Bq}$, and $r_{qC} = r_{Cq}$. The derivation of Eq. (5.41) was done for a three component mixture. A multi-component mixture of $n$ components is similarly described by

$$\frac{1}{T}\frac{\partial \mu_{i,T}}{\partial x} = \sum_{j\neq i}^{n} r_{ij}c_j(\mathbf{v}_i - \mathbf{v}_j) - r_{iq}J_q' \qquad \forall i = 1,\ldots,n-1 \quad (5.42)$$

$$\lambda\frac{\partial T}{\partial x} = \sum_{j\neq i}^{n} \frac{r_{jq}}{r_{qq}}c_j(\mathbf{v}_i - \mathbf{v}_j) - J_q' \qquad (5.43)$$

The thermal conductivity $\lambda = (r_{qq}T^2)^{-1}$ was introduced. Equations (5.42) and (5.43) can be used to eliminate the heat flux $J_q'$ and we obtain for all $i = 1,\ldots,n-1$

$$\frac{1}{T}\frac{\partial \mu_{i,T}}{\partial x} = \sum_{j\neq i}^{n} r_{ij}c_j(\mathbf{v}_i - \mathbf{v}_j) - \sum_{j\neq i}^{n} \frac{r_{iq}r_{jq}}{r_{qq}}c_j(\mathbf{v}_i - \mathbf{v}_j) + \lambda r_{iq}\frac{\partial T}{\partial x}$$

$$= \sum_{j\neq i}^{n} \left(r_{ij} - \frac{r_{iq}r_{jq}}{r_{qq}}\right)c_j(\mathbf{v}_i - \mathbf{v}_j) + \lambda r_{iq}\frac{\partial T}{\partial x} \qquad (5.44)$$

A symmetric set of Maxwell-Stefan diffusion coefficients can be defined analogous to earlier, see Section 5.1.3.

$$\left(r_{ij} - \frac{r_{iq}r_{jq}}{r_{qq}}\right) = -\frac{R}{c\mathcal{D}_{ij}} \qquad (5.45)$$

so that the transport due to a chemical potential gradient and a temperature gradient is

$$-\frac{1}{RT}\frac{\partial \mu_{i,T}}{\partial x} = \sum_{j\neq i}^{n} \frac{x_j}{\mathcal{D}_{ij}}(\mathbf{v}_i - \mathbf{v}_j) - \lambda\frac{r_{iq}}{R}\frac{\partial T}{\partial x} \qquad \forall i = 1,\ldots,n-1$$

$$(5.46)$$

Thermal diffusion is defined in the absence of chemical potential gradients. This motivates the relation

$$0 = \sum_{j\neq i}^{n} \frac{x_j}{\mathcal{D}_{ij}}(\mathbf{v}_i - \mathbf{v}_j) - \lambda\frac{r_{iq}}{R}\frac{\partial T}{\partial x} \qquad \forall i = 1,\ldots,n-1 \quad (5.47)$$

The velocity $\mathbf{v}_i$ is in this case solely due to thermal diffusion. From the definition $J_i = c_i\mathbf{v}_i$ and Eq. (4.19), we obtain the relation

$$\mathbf{v}_i = -D_i^T\frac{\partial T}{\partial x} \qquad (5.48)$$

where $D_i^T$ is the thermal diffusion coefficient, see Section 4.2. By introducing the thermal diffusion coefficient Eq. (5.48) in Eq. (5.47), we find for $i$ and $j$

$$0 = \sum_{j \neq i}^{n} \frac{x_j}{\mathcal{D}_{ij}} \left( D_i^T - D_j^T \right) + \lambda \frac{r_{iq}}{R} \qquad \forall i = 1, \ldots, n-1 \qquad (5.49)$$

and this can be reintroduced in Eq. (5.46) to finally give

$$-\frac{1}{RT} \frac{\partial \mu_{i,T}}{\partial x} = \sum_{j \neq i}^{n} \frac{x_j}{\mathcal{D}_{ij}} (v_i - v_j) + \sum_{j \neq i}^{n} \frac{x_j}{\mathcal{D}_{ij}} \left( D_i^T - D_j^T \right) \frac{\partial T}{\partial x} \qquad (5.50)$$

for $i = 1, \ldots, n-1$. Equation (5.49) can be seen as an alternative definition for the thermal diffusion coefficient.

## 5.3   Concluding remarks

The Maxwell-Stefan equations were at the center of interest in this Chapter, and their relation to other formulations of multi-component interdiffusion was shown. Their generalization to nonisothermal systems was pointed out. Predictions of multicomponent diffusion coefficients can now be done with high accuracy [62]. This gives a basis for numerous practical applications, in industrial systems (reactors, seperators) or in descriptions of natural processes.

Also coupled transports of mass can give work. Interdiffusion can, for instance, lead to work in biological systems. Thermal driving forces can lead to separation work, for example, in the presence of membranes, *cf.* Chapter 10.

# Chapter 6

# Systems with shear flow

*We derive the entropy production for a system with chemical reactions, temperature gradients and shear flow. The momentum balance contributes to a change in the internal energy. The Navier-Stokes equation and other relations are given. The Navier-Stokes equation contains a mechanical force due to the gradient of the reaction Gibbs energy. The rate of the chemical reaction obtains for symmetry reasons a term due to expansion or contraction. We discuss stationary pipe- and plug flow.*

Transport in pipes and other flow-equipment is central in chemical and mechanical engineering. Such flows exert viscous shear, which must be described by at least two coordinates. In Chapters 3–5 we assumed that the system was in mechanical equilibrium and that shear forces were absent. We now study viscous flow due to a pressure gradient. The second law of thermodynamics also governs this flow, and we shall see how the Navier-Stokes equation (the equation of motion) is related to the entropy production, and how the conjugate fluxes and forces can be defined in the presence of chemical reactions, temperature gradients and shear flow. These phenomena are typical in chemical reactors. The purpose of the Chapter is to give the central equations for a simple chemical reactor, or the stationary pipe- or plug flow reactor.

In order to find the entropy production, we again start with the time derivative of the Gibbs equation for a mixture of $n$ components, see also Appendix A.1

$$\frac{\partial s}{\partial t} = \frac{1}{T}\frac{\partial u}{\partial t} - \frac{1}{T}\sum_{j=1}^{n} \mu_j \frac{\partial c_j}{\partial t} \qquad (6.1)$$

Gibbs' equation is also valid for the spacial derivative and for the time derivative in a volume element that moves with the flow (the substantial time derivative). When the balance equations for entropy, mass, and internal energy are introduced into Eq. (6.1), we find the entropy flux as well as the entropy production, like we did in Chapter 3. The momentum balance was not needed in Chapter 3. We shall now include the momentum balance, and see that also viscous systems follow the structure given by Eqs. (1.1)–(1.3). Viscous contributions enter the Gibbs equation via the balance equation for internal energy.

## 6.1  Balance equations

The entropy balance, Eq. (3.1), for three-dimensional flow is

$$\frac{\partial s}{\partial t} = -\nabla \cdot \mathbf{J}_s + \sigma \qquad (6.2)$$

where nabla (the $\nabla \equiv (\partial/\partial x, \partial/\partial y, \partial/\partial z)$-operator) is a derivative -vector for all coordinate directions. The balance equations for mass, momentum, and internal energy are given below. More details on their derivations can be found in Appendix A.1.

### 6.1.1  Component balances

The mass balance for component $j$ is

$$\frac{\partial c_j}{\partial t} = -\nabla \cdot \mathbf{J}_j + \nu_j r \qquad (6.3)$$

where the molar flux vector is $\mathbf{J_j} = c_j v_j$, compare Eq. (3.2), and $c_j$ is given in mol·m$^{-3}$.

### 6.1.2  Momentum balance

The momentum balance, or equation of motion, for three-dimensional flow is

$$\frac{\partial \rho \mathbf{v}}{\partial t} = -\nabla \cdot (\rho \mathbf{v}\mathbf{v} + \mathbf{\Pi}) - \nabla p + \sum_{i=1}^{n} \rho_i \mathbf{f}_i \qquad (6.4)$$

where $\mathbf{\Pi}$ is the viscous pressure tensor.

### 6.1.3 Internal energy balance

The internal energy of a volume element changes with respect to time according to

$$\frac{\partial u}{\partial t} = -\nabla \cdot \left( \mathbf{J}_q' + \sum \mathbf{J}_i H_i \right) + \mathbf{v} \cdot \nabla p - \mathbf{\Pi} : \nabla \mathbf{v} \qquad (6.5)$$

See also Appendix A.1. The first term on the right-hand side is the divergence of the total heat flux in an electroneutral system, see Chapter 3. The second term adds energy to the volume element from changes in pressure, while the last term adds internal energy because of shear. The viscous pressure tensor $\mathbf{\Pi}$ is referred to in Fluid Mechanics as the deviatoric stress tensor, with a negative sign. The last term is called the Rayleigh dissipation function, see e.g. [2]. The viscous flow term in Eq. (6.5) leads to an increasing internal energy in the volume element, just like a heat flux could. The contribution is thus sometimes referred to as energy dissipated as heat. Viscous flow leads to entropy production, as we shall see in the exercise below.

## 6.2 Entropy production

Consider a volume element moving with a flow. The center-of-mass, or barycentric, velocity is $\mathbf{v}$. This velocity is defined by $\mathbf{v} \equiv \frac{1}{\rho} \sum_{i=1}^{n} \mathbf{v}_i \rho_i = \sum_{i=1}^{n} \mathbf{v}_i w_i$ where $w_i = \rho_i/\rho$ is the mass-fraction of component $i$. Diffusion of a component with respect to the barycentric frame of reference is

$$\mathbf{J}_{i,\text{bar}} \equiv c_i (\mathbf{v}_i - \mathbf{v}) \qquad (6.6)$$

while the flux of the same component with respect to the wall is

$$\mathbf{J}_i = c_i \mathbf{v}_i = c_i \mathbf{v} + \mathbf{J}_{i,\text{bar}} \qquad (6.7)$$

The second-last term is called the convective and the last part the diffusive part of the total flux. Using the fact that $\rho_i = M_i c_i$ it follows from the definition of $\mathbf{J}_{i,\text{bar}}$ together with the mass balance that

$$\sum_{i=1}^{n} M_i \mathbf{J}_{i,\text{bar}} = 0 \qquad (6.8)$$

so the barycentric diffusion fluxes are not independent of one another. The entropy balance is

$$\frac{\partial s}{\partial t} = -\nabla \cdot \mathbf{J}_s + \sigma$$

$$= -\nabla \cdot \left( \frac{\mathbf{J}'_q}{T} + s\mathbf{v} + \mathbf{J}_{s,\mathrm{bar}} \right) + \sigma \tag{6.9}$$

The total entropy flux, $\mathbf{J}_s$, has three contributions. The second line distinguishes between the entropy flux due to the measurable heat flux and the entropy carried by the mass flux across the system boundary. Using Eq. (6.7), the entropy flux due to mass flux can further be decomposed into a convective and a diffusive term: The entropy density of the mixture, $s = \sum_i c_i S_i$, carried by the convective flow and the entropy carried by diffusion with respect to the center of mass, $\mathbf{J}_{s,\mathrm{bar}} = \sum_i \mathbf{J}_{i,\mathrm{bar}} S_i$, where $S_i$ is the partial molar entropy. The entropy production in the volume element, $\sigma$, is positive according to the second law of thermodynamics. In order to obtain explicit expressions for the entropy production, we introduce the component balance (Eq. 6.3) and the balance for internal energy (Eq. 6.5) in the Gibbs equation Eq. (6.1). This gives

$$\frac{\partial s}{\partial t} = -\frac{1}{T} \nabla \cdot \left( \mathbf{J}'_q + \sum_{i=1}^{n} \mathbf{J}_i H_i \right) + \frac{1}{T} \mathbf{v} \cdot \nabla p - \frac{1}{T} \mathbf{\Pi} : \nabla \mathbf{v}$$

$$+ \sum_{i=1}^{n} \frac{\mu_i}{T} \nabla \cdot \mathbf{J}_i - r \frac{\Delta_r G}{T} \tag{6.10}$$

This equation can rewritten as

$$\frac{\partial s}{\partial t} = -\nabla \cdot \left( \frac{\sum_{i=1}^{n} \mathbf{J}_i (H_i - \mu_i) + \mathbf{J}'_q}{T} \right) + \mathbf{J}'_q \cdot \nabla \frac{1}{T} + \sum_{i=1}^{n} \mathbf{J}_i H_i \cdot \nabla \frac{1}{T}$$

$$- \sum_{i=1}^{n} \mathbf{J}_i \cdot \nabla \frac{\mu_i}{T} + \frac{1}{T} \mathbf{v} \cdot \nabla p - \frac{1}{T} \mathbf{\Pi} : \nabla \mathbf{v} - r \frac{\Delta_r G}{T} \tag{6.11}$$

By comparing Eq. (6.11) to Eq. (6.9), we confirm the total entropy flux $\mathbf{J}_s$ according to the second line of Eq. (6.9), and we identify the entropy production as

$$\sigma = \mathbf{J}'_q \cdot \nabla \frac{1}{T} + \sum_{i=1}^{n} \mathbf{J}_i H_i \cdot \nabla \frac{1}{T} - \sum_{i=1}^{n} \mathbf{J}_i \cdot \nabla \frac{\mu_i}{T} + \frac{1}{T} \mathbf{v} \cdot \nabla p$$

$$- \frac{1}{T} \, \mathbf{\Pi} : \nabla \mathbf{v} - r \, \frac{\Delta_r G}{T}$$

$$= \mathbf{J}'_q \cdot \nabla \frac{1}{T} - \frac{1}{T} \sum_{i=1}^{n} \mathbf{J}_i \cdot (S_i \nabla T + \nabla \mu_i) + \frac{1}{T} \mathbf{v} \cdot \nabla p$$

$$- \frac{1}{T} \, \mathbf{\Pi} : \nabla \mathbf{v} - r \, \frac{\Delta_r G}{T} \tag{6.12}$$

We use Gibbs-Duhem's equation to eliminate terms proportional to $\mathbf{v}$, and the definition of the gradient of the chemical potential in the direction of constant temperature $\nabla \mu_{i,T} = \nabla \mu_i + S_i \nabla T$ (see Appendix Eq. (A.66)). The entropy production then obtains a convenient form

$$\sigma = \mathbf{J}'_q \left( \nabla \frac{1}{T} \right) + \sum_{i=1}^{n} \mathbf{J}_{i,\text{bar}} \left( -\frac{1}{T} \nabla \mu_{i,T} \right) + \mathbf{\Pi} : \left( -\frac{1}{T} \nabla \mathbf{v} \right) + r \left( -\frac{\Delta_r G}{T} \right) \tag{6.13}$$

The entropy production is a sum of flux-force products. The fluxes are the measurable heat flux, the diffusional fluxes, the viscous pressure tensor and the rate of the chemical reaction. The forces that conjugate to these are the gradient in the inverse temperature, minus the gradient in the chemical potential in the direction of constant temperature over the temperature, minus the gradient in the velocity over the temperature and minus the reaction Gibbs energy over the temperature. All flux-force pairs contribute to the entropy production and to the dissipation of energy as heat. If another velocity other than the barycentric velocity is chosen as a frame of reference, the above equations will be modified, see Eqs. (6.22) and (6.23) below, but the value of $\sigma$ remains the same.

The first two contributions to $\sigma$ are vectorial. The third term is the product of two tensors of rank two. The last term contains scalars (tensors of rank zero). To describe the force-flux relation between gradient in velocity and viscous pressure tensor, we analyze both tensors. The gradient in velocity can be decomposed into three contributions

$$\nabla \mathbf{v} = \underbrace{\frac{1}{2}\left(\nabla \mathbf{v} + (\nabla \mathbf{v})^T\right) - \frac{1}{3}(\nabla \cdot \mathbf{v})\mathbf{1}}_{\text{rate-of-shear tensor}} + \underbrace{\frac{1}{3}(\nabla \cdot \mathbf{v})\mathbf{1}}_{\text{rate of volume expansion}}$$

$$+ \underbrace{\frac{1}{2}\left(\nabla \mathbf{v} - (\nabla \mathbf{v})^T\right)}_{\text{rate of rotation}} \tag{6.14}$$

Superscript T indicates the transpose of a matrix, and the divergence of the velocity is $\nabla \cdot \mathbf{v} \equiv \partial v_x/\partial x + \partial v_y/\partial y + \partial v_z/\partial z$. The first contribution is a symmetric tensor that describes the shape deformation of a fluid element. The second term is a diagonal tensor capturing the expansion of the volume element. The last antisymmetric tensor describes a rigid-like rotation. Multiplying the rotation tensor with the viscous pressure tensor $\boldsymbol{\Pi}$ gives zero, noting that $\boldsymbol{\Pi}$ is symmetric. The rigid-like rotation of the fluid does not produce entropy. The rotation does not contribute to the deformation of the fluid and it is not part of the viscous force-flux relation.

The viscous pressure tensor is not traceless, although some Fluid Mechanics texts implicitly make this assumption. The viscous pressure tensor can be written as the sum of a symmetric traceless tensor of rank two, $\overline{\overline{\boldsymbol{\Pi}}}$, and the trace which is a scalar, as

$$\boldsymbol{\Pi} = \overline{\overline{\boldsymbol{\Pi}}} + \frac{1}{3}\,\mathbf{1}\,\mathrm{Tr}(\boldsymbol{\Pi}) \tag{6.15}$$

where Tr denotes the trace. The entropy production can now be written as

$$\sigma = \sigma_{\text{vect}} + \sigma_{\text{tens}} + \sigma_{\text{scal}} \tag{6.16}$$

where

$$\sigma_{\text{vect}} = \mathbf{J}'_q \cdot \nabla \frac{1}{T} - \sum_{i=1}^{n} \mathbf{J}_{i,\text{bar}} \cdot \frac{1}{T} \nabla \mu_{i,T}$$

$$= \mathbf{J}'_q \cdot \nabla \frac{1}{T} - \sum_{i=1}^{n-1} \mathbf{J}_{i,\text{bar}} \cdot \frac{1}{T} \nabla \left(\mu_{i,T} - \frac{M_i}{M_n}\mu_{n,T}\right)$$

$$\sigma_{\text{tens}} = -\frac{1}{T}\,\overline{\overline{\boldsymbol{\Pi}}} : \left(\frac{1}{2}\left(\nabla \mathbf{v} + (\nabla \mathbf{v})^T\right) - \frac{1}{3}\left(\nabla \cdot \mathbf{v}\right)\mathbf{1}\right)$$

$$\sigma_{\text{scal}} = -\frac{1}{3T}\,\mathrm{Tr}\left(\boldsymbol{\Pi}\right)\nabla \cdot \mathbf{v} - r\frac{\Delta_r G}{T} \tag{6.17}$$

We used $\overline{\overline{\Pi}} : \mathbf{1} = 0$, realizing that $\overline{\overline{\Pi}}$ is traceless, and analogously the product of the traceless rate-of-shear tensor, *cf.* Eq. (6.14), with $\mathbf{1}$ vanishes. Coupling takes place only between tensors of the same order (the Curie principle). The vectorial contribution leads to linear force-flux relations that have already been discussed in the previous Chapter, where we started with expressions for the forces in terms of the fluxes. Using the Gibbs-Duhem equation, it was then possible to show that the equation for $\nabla \mu_{n,T}$ followed from the equations for $\nabla \mu_{i,T}$, $i = \overline{1, n-1}$. In these equations all velocities could be replaced by velocity differences. The diffusion fluxes contain velocity differences with the barycentric velocity. As a consequence $\mathbf{J}_{n,\mathrm{bar}}$ is a function of the other diffusion fluxes. In Eq. (6.17a) we have used Eq. (6.8) to eliminate $\mathbf{J}_{n,\mathrm{bar}}$. This procedure is an alternative way of incorporating the Gibbs-Duhem equation into the one used in Chapter 5.

In the tensorial contribution to the entropy production there is only one force-flux pair and the resulting linear law is

$$\overline{\overline{\Pi}} = -\eta \left( \nabla \mathbf{v} + (\nabla \mathbf{v})^{\mathrm{T}} - \frac{2}{3} \left( \nabla \cdot \mathbf{v} \right) \mathbf{1} \right) \qquad (6.18)$$

where the coefficient $\eta$ is the shear viscosity. A factor $1/(2T)$ was absorbed into $\eta$ in Eq. (6.18). The factor $\frac{1}{2}$ leads to a more intuitive definition of $\eta$ because experiments are designed such that only one off-diagonal element of the velocity gradient contributes to the bracket in Eq. (6.18) (*cf.* exercise below). The so-defined shear viscosity coincides with Newton's law of friction.

In the scalar contribution to the entropy production there are two force-flux pairs, and the resulting linear equations are

$$\frac{1}{3} \mathrm{Tr} \left( \mathbf{\Pi} \right) = -\zeta \, \nabla \cdot \mathbf{v} - \lambda_r \, \Delta_r G$$
$$r = -\lambda_r \, \nabla \cdot \mathbf{v} - L_r \, \Delta_r G \qquad (6.19)$$

where $\zeta$ is the bulk viscosity (or second viscosity) and $\lambda_r$ is the chemical viscosity. Note, that $\nabla \cdot \mathbf{v} = \mathrm{Tr} \left( \nabla \mathbf{v} \right)$, so that the trace of the velocity gradient is the conjugate driving force to the trace of the viscous pressure tensor. A factor $1/(3T)$ was absorbed into $\zeta$ in Eqs. (6.19). The viscosities are functions of temperature, density, and composition. The shear and the bulk viscosities have dimension Pa·s, and the chemical viscosity has dimension $m^{-3}$. The total viscous pressure tensor, or the law of Navier-Poisson, is

$$\mathbf{\Pi} = -\eta \left( \nabla \mathbf{v} + (\nabla \mathbf{v})^\mathrm{T} - \frac{2}{3} \left( \nabla \cdot \mathbf{v} \right) \mathbf{1} \right) - \zeta \left( \nabla \cdot \mathbf{v} \right) \mathbf{1} - \lambda_r \Delta_r G \, \mathbf{1} \quad (6.20)$$

The second term on the right-hand side in Eq. (6.20) captures the viscous pressure contribution of an expanding or contracting fluid flow. The contribution from the second term is zero for incompressible fluids, where $\nabla \cdot \mathbf{v} = 0$. It also disappears for monoatomic fluids, then the bulk viscosity $\zeta$ is zero. For molecular fluids undergoing a rapid volume-expansion the term contributes to the viscous pressure tensor, Eq. (6.20), and to the entropy production, Eq. (6.19).

The third term describes the coupling of the tensor to a chemical reaction. Meixner [65] explained that the chemical viscosity is caused by a chemical reaction which progresses quickly compared to the rate of exchange of momentum. The third term is usually not taken into account.

Eliminating all but one non-diagonal viscous pressure tensor element leads to Newton's law of friction, see Eq. (2.5). That equation describes laminar flow in the $x$-direction, with velocity $\mathbf{v} = (\mathrm{v}_x, 0, 0)$ and velocity component $\mathrm{v}_x = \mathrm{v}_x(y)$. Experiments show that the shear viscosity of non-Newtonian fluids is a function of the shear, $\nabla \mathbf{v}$. The structure of Eqs. (1.1)–(1.3), i.e. the linear force-flux relation, is then not preserved, because the viscosity should not depend on a thermodynamic force. It is possible to generalize non-equilibrium thermodynamic theory to include this case, by introducing the orientation of the molecules as an internal variable [66, 67, 68], see Chapter 7.

Navier-Stokes equation is obtained by substituting the viscous pressure tensor, Eq. (6.20), into the equation of motion (noting that $\nabla \cdot (\nabla \mathbf{v})^\mathrm{T} = \nabla (\nabla \cdot \mathbf{v})$), assuming $\eta$, $\zeta$ and $\lambda_r$ constant, as

$$\frac{\partial \rho \mathbf{v}}{\partial t} = - \nabla \cdot \rho \mathbf{v} \mathbf{v} + \eta \, \Delta \mathbf{v} + \left( \zeta + \frac{1}{3} \eta \right) \nabla \nabla \cdot \mathbf{v}$$

$$+ \lambda_r \nabla \Delta_r G - \nabla p + \sum_{i=1}^{n} \rho_i \mathbf{f}_i \quad (6.21)$$

where $\Delta \equiv \nabla \cdot \nabla$ is the Laplace operator. A force term appears in the Navier-Stokes equation due to a gradient of the reaction Gibbs energy. By substituting Eq. (6.19) for the reaction rate into the component balance equation, Eq. (6.3), we obtain

$$\frac{\partial c_j}{\partial t} = -\nabla \cdot \mathbf{J}_j - \nu_j \lambda_r \, \nabla \cdot \mathbf{v} - \nu_j L_r \Delta_r G$$

This shows that expansion and compression gives a contribution to the reaction rate and a corresponding change of the concentrations. As we discussed above, the contribution containing $\lambda_r$ is usually neglected.

The vectorial fluxes, tensors of rank one, can couple to one another, but not to the tensors of rank zero or two. No other terms can couple to the shear viscosity. But the bulk viscosity term may couple to the chemical reaction rate. This effect has so far not been investigated in detail.

The entropy production is independent of the frame of reference, but several fluxes are not. The flux, $\mathbf{J}_{i,\text{bar}}$, of component $i$ relative to the barycentric velocity was introduced in the reformulation of Eq. (6.12) into Eq. (6.13). If a different velocity field, $\mathbf{v}_{\text{ref}}(\mathbf{r}, t)$, is chosen as the frame of reference, the component fluxes become

$$\mathbf{J}_i = c_i \mathbf{v}_{\text{ref}} + \mathbf{J}_{i,\text{ref}} \tag{6.22}$$

$$\mathbf{J}_s = \frac{\mathbf{J}_q'}{T} + s\mathbf{v}_{\text{ref}} + \sum_{i=1}^{n} \mathbf{J}_{i,\text{ref}}\, S_i$$

The corresponding entropy production is

$$\sigma = \mathbf{J}_q' \cdot \nabla \frac{1}{T} - \sum_{i=1}^{n} \mathbf{J}_{i,\text{ref}} \cdot \frac{1}{T} \nabla \mu_{i,T}$$

$$+ (\mathbf{v} - \mathbf{v}_{\text{ref}}) \cdot \frac{1}{T} \nabla p - \frac{1}{T} \mathbf{\Pi} : \nabla \mathbf{v} - r \frac{\Delta_r G}{T} \tag{6.23}$$

where $(\mathbf{v} - \mathbf{v}_{\text{ref}}) \cdot (\nabla p)/T$ appears as an additional vectorial flux-force pair. From the discussion in the previous Section it is clear that two of the vectorial flux-force pairs depend on the others. This makes the use of an arbitrary reference velocity impractical when $\nabla p \neq 0$.

The term disappears when the barycentric velocity is chosen as the frame of reference, $\mathbf{v}_{\text{ref}} = \mathbf{v}$. This is the case discussed in the previous Section.

**Exercise 6.2.1** *Consider a fluid in stationary shear flow between two plates of infinite length positioned at $z = 0$ and $z = z_0$. The lower plate is fixed while the upper plate moves with velocity $v_{x0}$ in the $x$-direction. The configuration can be used to measure a fluid's viscosity. The upper and lower temperatures are $T_u$ and $T_l$, respectively. Assume that the viscosity, $\eta$, and thermal conductivity, $\lambda$, are constant. Calculate the velocity profile and the temperature profile.*

- **Solution:** We want to calculate the two fields $v_x = v_x(z)$ and $T = T(z)$ with the four boundary conditions $v_x(z = 0) = 0$, $v_x(z = z_0) = v_{x0}$, $T(z = 0) = T_l$, and $T(z = z_0) = T_u$. As there is no pressure difference along the plates the pressure $p$ is constant everywhere. The momentum balance for the velocity field is then

$$\frac{\mathrm{d}}{\mathrm{d}z}\left(\Pi_{xz}\right) = 0 \quad \Rightarrow \quad -\eta\frac{\mathrm{d}^2 v_x}{\mathrm{d}z^2} = 0$$

$$\Rightarrow \quad v_x = c_1 z + c_2 \quad \Rightarrow \quad v_x = v_{x0}\frac{z}{z_0}$$

The internal energy balance for the temperature field gives

$$-\nabla\cdot\mathbf{J}'_q = \mathbf{\Pi} : \nabla\mathbf{v} \quad \Rightarrow \quad -\lambda\frac{\mathrm{d}^2 T}{\mathrm{d}z^2} = \eta\left(\frac{\mathrm{d}v_x}{\mathrm{d}z}\right)^2$$

$$\Rightarrow \quad -\lambda\frac{\mathrm{d}^2 T}{\mathrm{d}z^2} = \eta\left(\frac{v_{x0}}{z_0}\right)^2$$

$$\Rightarrow \quad T(z) = -\frac{\eta}{\lambda}\left(\frac{v_{x0}}{z_0}\right)^2 z^2 + c_3 z + c_4$$

and

$$T(z) = -\frac{\eta}{\lambda}v_{x0}\left(\frac{z}{z_0}\right)^2 + \left((T_u - T_l) + \frac{\eta}{\lambda}v_{x0}\right)\left(\frac{z}{z_0}\right) + T_l$$

The four integration constants $c_1, \ldots c_4$ were determined from the four boundary conditions. The result is a velocity profile which is linear between the two plates, and a temperature profile which is parabolic, see Fig. 6.1. The viscous shear leads to a heat flux from the fluid in both directions to the plates.

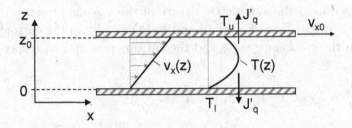

Figure 6.1: Velocity and temperature profile for a shear flow between two plates.

The exercise shows that a heat flux $-\lambda(\mathrm{d}T/\mathrm{d}z)$ is due to viscous flow. The heat flux is a sign of the irreversibility of the shear flow. The entropy flux into the surroundings leads to lost work according to Section 2.3.

## 6.3   Stationary pipe flow

Material is flowing along a pipe because a pressure difference is applied. For many practical problems it is appropriate to consider stationary flow conditions. The mechanical forces are balanced in the stationary state. It follows from Eq. (A.14) or from Eq. (6.21) that

$$0 = \nabla \cdot (\rho \mathbf{v}\mathbf{v} + \mathbf{\Pi} + p\mathbf{1}) \tag{6.24}$$

We restrict ourselves to incompressible laminar flow, where $\rho$ is constant and $\nabla \cdot \mathbf{v} = 0$. Eq. (6.24) then simplifies to

$$\nabla \cdot \mathbf{\Pi} = -\nabla p \tag{6.25}$$

The entropy production Eq. (6.13) contains a term $1/T\,\mathbf{\Pi} : \nabla \mathbf{v}$ that can be separated into three parts. As the viscous pressure is now controlled by the pressure gradient, rather than being an independent response to the velocity gradient, we should rewrite the last two equations to reflect this. We have

$$\frac{1}{T}\mathbf{\Pi} : \nabla \mathbf{v} = \nabla \cdot \frac{\mathbf{\Pi} \cdot \mathbf{v}}{T} - (\nabla \cdot \mathbf{\Pi}) \cdot \frac{\mathbf{v}}{T} - \mathbf{v} \cdot \mathbf{\Pi} \cdot \nabla \frac{1}{T}$$

$$= \nabla \cdot \frac{\mathbf{\Pi} \cdot \mathbf{v}}{T} + \frac{\mathbf{v}}{T} \cdot \nabla p - \mathbf{v} \cdot \mathbf{\Pi} \cdot \nabla \frac{1}{T} \tag{6.26}$$

where we used the symmetric nature of the viscous pressure tensor, and Eq. (6.25). From Eq. (6.9) and (6.13) we obtain the entropy balance in the stationary state and the entropy production, respectively

$$\nabla \cdot \left( \frac{\mathbf{J}_q'}{T} + s\mathbf{v} + \sum_i \mathbf{J}_{i,\text{bar}} S_i \right) = \sigma$$

$$= (\mathbf{J}_q' + \mathbf{v} \cdot \mathbf{\Pi}) \cdot \nabla \frac{1}{T} - \sum_{i=1}^{n} \mathbf{J}_{i,\text{bar}} \cdot \frac{\nabla \mu_{i,T}}{T}$$

$$- r \frac{\Delta_r G}{T} - \frac{\mathbf{v}}{T} \cdot \nabla p - \nabla \cdot \frac{\mathbf{\Pi} \cdot \mathbf{v}}{T} \quad (6.27)$$

The last term on the right-hand side of this equation can be grouped with the flux term on the left. We then observe the structure of the balance equation as

$$\nabla \cdot \left( \frac{\mathbf{J}_{q,\text{pipe}}'}{T} + s\mathbf{v} + \sum_i \mathbf{J}_{i,\text{bar}} S_i \right) = \sigma_{\text{pipe}} \quad (6.28)$$

with

$$\sigma_{\text{pipe}} = \mathbf{J}_{q,\text{pipe}}' \cdot \nabla \frac{1}{T} - \sum_{i=1}^{n} \mathbf{J}_{i,\text{bar}} \cdot \frac{\nabla \mu_{i,T}}{T} - \frac{\mathbf{v}}{T} \cdot \nabla p - r \frac{\Delta_r G}{T} \quad (6.29)$$

A new heat flux appears

$$\mathbf{J}_{q,\text{pipe}}' = \mathbf{J}_q' + \mathbf{v} \cdot \mathbf{\Pi} \quad (6.30)$$

Here the name "energy flux", may be more appropriate. The condition of mechanical equilibrium in the pipe flow has led to two important changes in the equations. The first is that the viscous force-flux pair in the entropy is replaced by a velocity $(\nabla p) / T$ flux-force pair. This flux-force pair gives Darcy's law for pipe flow. The second change is an additional $\mathbf{v} \cdot \mathbf{\Pi}$ contribution in the energy flux. Experimental or theoretical studies of pipe flow should take this into account. The additional contribution to the energy flux, appears in the energy balance.

The resulting flux-force relations for the vectorial fluxes are

$$\mathbf{J}'_{q,\text{pipe}} = L_{qq} \nabla \frac{1}{T} - \frac{1}{T} \sum_{i=1}^{n} L_{qi} \nabla \mu_{i,T} - L_{qp} \frac{1}{T} \nabla p$$

$$\mathbf{J}_{j,\text{bar}} = L_{jq} \nabla \frac{1}{T} - \frac{1}{T} \sum_{i=1}^{n} L_{ji} \nabla \mu_{i,T} - L_{jp} \frac{1}{T} \nabla p$$

$$\mathbf{v} = L_{pq} \nabla \frac{1}{T} - \frac{1}{T} \sum_{i=1}^{n} L_{pi} \nabla \mu_{i,T} - L_{pp} \frac{1}{T} \nabla p \qquad (6.31)$$

The matrix of conductivities is symmetric. The last equation is Darcy's law, relating the fluid velocity to the pressure gradient. For porous media, the coefficient $L_{pp}$ is found to scale with $L_{pp}/T = \kappa/\eta$, i.e. with the inverse of viscosity and with permeability $\kappa$. For the scalar flux-force pair one obtains

$$r = -L_r \Delta_r G \qquad (6.32)$$

where we absorbed the factor $1/T$ in $L_r$.

## 6.4 The plug flow reactor

The plug flow model for chemical reactions gives a first order approximation to phenomena that take place in a chemical reactor. Equation (6.29) is used as a starting point. Thermal conduction is frequently neglected in the axial direction. Diffusion relative to the barycentric frame of reference is also neglected, and a pressure gradient is assumed only in axial direction, $z$, of the reactor, not in the radial direction. The local entropy production becomes

$$\sigma = J'_q(r) \frac{\mathrm{d}}{\mathrm{d}r} \frac{1}{T(r)} - \frac{\mathrm{v}}{T} \frac{\mathrm{d}p(z)}{\mathrm{d}z} - r \frac{\Delta_r G}{T} \qquad (6.33)$$

where $J'_q$ is the measurable heat flux in radial direction, and v is the velocity in the $z$-direction. We integrate over the cross section ($\int ...2\pi r \, \mathrm{d}r$) and obtain the entropy production per unit length

$$\dot{\sigma}' = \Omega \, r \left( -\frac{\Delta_r G}{T} \right) + \pi D \, J'_q(r = D) \, \Delta_{\text{shell}} \frac{1}{T} + \Omega \, \mathrm{v} \left( -\frac{1}{T} \frac{\mathrm{d}p}{\mathrm{d}z} \right) \quad (6.34)$$

where $(r \, J'_q(r))$ is constant in $r$-direction across the tube-shell as required from the energy balance for steady state conditions. The tube has the outside diameter $D$. We introduced $\Omega$ as the inner cross-sectional area of the tube. The difference of inverse temperature, $\Delta_{shell}(1/T)$, is taken from the outside of the tube-shell to the inside. This equation is the starting point of the studies of energy efficient reactor design in Section 11.4.2, see Eq. (11.33).

## 6.5 Transport coefficients: viscosity and thermal conductivity

In order to model the system, we need information on the transport coefficients. The main coefficients are the viscosity, the thermal conductivity, and kinetic coefficients related to chemical reactions. The last are described in chapter 7. For low density mixtures, coefficients can be obtained from kinetic gas theory. In other cases, the transport coefficients must be measured or estimated from molecular simulations [69] or group contribution methods [70, 71]. Force fields are needed as input for molecular simulations.

Rosenfeld already showed in 1977 [72, 73], that for pure, simple fluids, the dimensionless viscosity $\eta^*(s^{res})$, the thermal conductivity $\lambda^*(s^{res})$, and the self-diffusion coefficient $D^{*self}(s^{res})$ depended on the residual entropy, $s^{res}$, only. The residual entropy is the difference of the entropy and the ideal gas entropy, $s^{res} = s - s^{ig}$. The asterisk * indicates that the quantity is dimensionless. It is the ratio of the actual value and a reference value. The proper choice of the reference value is subject to ongoing work. For viscosities, very good results have been obtained using the Chapman-Enskog theory[74] as a reference, *i.e.* $\eta* = \eta/\eta_{CE}$ [75].

The dependence on $s^{res}$ was recently confirmed to hold also for non-spherical compounds [76, 75] and for strongly hydrogen-bonding substances, such as water [71]. Lötgering-Lin and Gross [71] developed a predictive method based on a group-contribution approach using the PC-SAFT equation of state [77]. The residual entropy is then obtained analytically as a derivative of the Helmholtz energy, as described in Refs. [77, 70]. The results are shown in Fig. 6.3.

Figure 6.2 presents the experimental viscosity of n-octane as a function of temperature and pressure. Figure 6.3 shows the same experimental viscosity data versus the residual entropy as calculated from

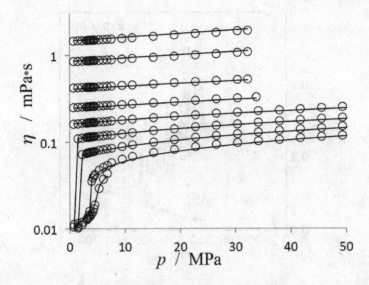

Figure 6.2: Viscosity of n-octane (for $T = 149$ K to 442 K). Comparison of experimental data (symbols) with results from entropy scaling using PC-SAFT model for residual entropy (lines).

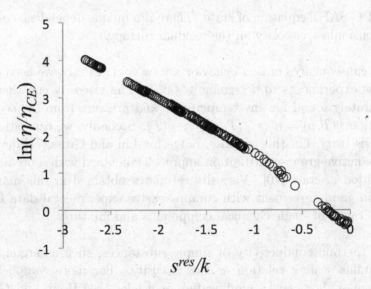

Figure 6.3: Dimensionless viscosity $\eta^*$ of n-octane versus residual entropy computed from the PC-SAFT equation of state, according to Lötgering-Lin and Gross [71].

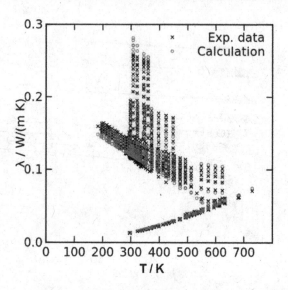

Figure 6.4: Thermal conductivity of n-hexane for various pressures. Comparison of experimental data with results from entropy scaling using PC-SAFT model for residual entropy.

the PC-SAFT equation of state. There is a unique dependence of the dimensionless viscosity on the residual entropy.

The consequences of this behavior are twofold. Firstly, we need only a few experiments to determine $\eta^*(s^{res})$. The viscosity can then be accurately found for any temperature and pressure from the scaling relation $\eta(T,p) = \eta^*(s^{res}(T,p)) \cdot \eta_{CE}(T)$. Secondly, we can estimate the viscosity. For this purpose, Lötgering-Lin and Gross [71] devised a predictive group-contribution approach combined with a predictive equation of state [70]. Viscosity estimates obtained in this manner are in good agreement with comprehensive experimental data for a wide range of single chemical compounds and mixtures.

The thermal conductivity of simple substances, such as argon, has a similar scaling relation as the viscosity. For more complicated molecules, however, a modification is needed, see Hopp and Gross [78]. The dimensionless thermal conductivity is then written in the form $\lambda^*(s^{res}) = \lambda/(\lambda_{CE} + \alpha(s^{res}) \cdot \lambda_{vib})$, where $\alpha(s^{res})$ is a function of the residual entropy and $\lambda_{vib}$ is the vibrational contribution to the

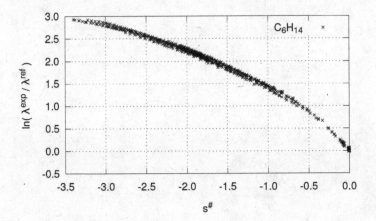

Figure 6.5:  Dimensionless thermal conductivity $\lambda^* = \lambda/(\lambda_{CE} + \alpha(s^{res})\lambda_{vib})$ of n-hexane versus residual entropy from PC-SAFT model, according to Hopp and Gross [78].

thermal conductivity. Using this revised expression, the scaling relation holds for wide ranges of conditions. The data in Fig. 6.4 can then be much simpler pictured, as shown in Fig. 6.5.

## 6.6  Concluding remarks

We have seen that the systematic treatment prescribed by non-equilibrium thermodynamics helps to define and relate fluxes and forces in systems with viscous flow. The Navier-Stokes equation fits well into the format of the theory. The entropy production is the origin of viscous heating or Rayleigh dissipation. Coupling of the reaction to compressional shearing is little studied.

Our analysis allows a systematic evaluation of the shear contribution to the entropy production and the equations of motion. This has a bearing on the optimal design of process equipment. The optimal state may arise from a trade-off between shear contributions (viscous dissipation) and losses due to heat and mass transport, see Chapter 11.

# Chapter 7

# Chemical reactions

*Chemical reactions have rates which normally are non-linear in their driving force. They are described by the law of mass action. In order to derive the law of mass action from non-equilibrium thermodynamics, we shall introduce an internal variable. The internal variable is the probability that the reaction is in a state $\gamma$ on the path between the reactant and product states. The process along this path occurs on a mesoscopic level. The extended theory is called mesoscopic non-equilibrium thermodynamics. A linear force-flux relation along the reaction path leads to the law of mass action. Mesoscopic non-equilibrium thermodynamics captures the nature of the chemical reaction, and gives it a thermodynamic basis.*

Spontaneous chemical reactions have high rates of energy dissipation. The contribution to the entropy production from a chemical reaction was given in Chapter 3

$$\sigma = r\left(-\frac{\Delta_r G}{T}\right) \tag{7.1}$$

We shall learn in Chapter 11 how this entropy production can be minimized in chemical reactors. The expression (7.1) is valid, whenever the Gibbs equation is valid, *cf.* remark 2 in Chapter 3.

The relation between the reaction rate $r$ and $\Delta_r G$ is normally non-linear. In the theory of non-equilibrium thermodynamics discussed in Chapters 1–6, the fluxes (rates) are linear functions of the driving forces. For chemical reactions we then have

$$r = -l\frac{\Delta_r G}{T} \qquad (7.2)$$

For a numerical example, see [79]. But such relations do not normally apply to chemical reactions. Take as an example the elementary reaction

$$A + B \rightleftharpoons C + D$$

In standard chemical reaction kinetics [80] the rate is described by the law of mass action:

$$r = k_f c_A c_B - k_r c_C c_D \qquad (7.3)$$

where $k_f$ and $k_r$ are kinetic constants for the forward and the reverse reaction, and $c_i$ is the concentration of $i$.

A rate like Eq. (7.3) can be used with Eq. (7.1), but a thermodynamic expression for $r$ which contains the explicit limit that $r = 0$ for $\Delta_r G = 0$, is preferable for instance in optimization studies. The purpose of this Chapter is to present such an expression. We shall see that the thermodynamic form contains a generalized version of the law of mass action. The thermodynamic expression for the reaction rate is useful for engineers who want to consider a chemical reactor, not only as a producer of chemicals, but also as a work-producing or work-consuming apparatus. This viewpoint is essential for designing an optimal reactor with minimal entropy production, where fluxes of heat, mass, and reaction rate have to be balanced, *cf.* Chapter 11.

The variables of the Gibbs energy $\Delta_r G(T, p, N_1, ...N_n)$ of Eqs. (7.1) and (7.2) are macroscopic variables, meaning that they can be controlled from the outside of the system. The Gibbs energy of reaction is determined by two states: the reactant state and the product state. We need to describe the transition from reactants to products on a more detailed level to find the rate law. For this purpose we introduce the reaction path along which the reacting complex passes on its way from reactants to products. This reaction path, which is a well-known concept in reaction kinetics, will also be refered to as the internal coordinate for the chemical reaction. We describe it with the continuous coordinate $\gamma$, which is chosen to have a value between 0 (reactants) and 1 (products). This choice can be done without any loss of generality. The probability $c(\gamma)$ that the reaction is in state $\gamma$ is the internal variable. It is not possible to independently control an internal variable from the outside.

The use of internal variables was first proposed by Prigogine and Mazur, see Ref. [12]. They have been used by Rubi and coworkers [81, 28, 82] to describe nucleation, evaporation, and molecular pumps.

The extension of the theory to the mesoscopic level is called *mesocopic* non-equilibrium thermodynamics. In this Chapter, we shall see how mesocopic non-equilibrium thermodynamics can be used to give rate equations that are useful in thermodynamic applications as well as in reaction kinetics. We first recapitulate the normal procedure to calculate the chemical driving force $-\Delta_r G/T$ in Sec. 7.1 before we proceed to the mesoscopic description in Sec. 7.2. The temperature shall be taken as constant in the analysis in this Chapter.

## 7.1  The Gibbs energy change of a chemical reaction

The variation in the Gibbs energy with the composition, at constant $p, T$, for the above reaction is

$$dG = \mu_A dN_A + \mu_B dN_B + \mu_C dN_C + \mu_D dN_D \qquad (7.4)$$

The Gibbs energy of a mixture is given in Appendix A.2. The reaction relates the changes in mole numbers, leading to the definition of the extent of the reaction, $d\xi$

$$d\xi \equiv dN_C = dN_D = -dN_A = -dN_B \qquad (7.5)$$

In general, we have

$$d\xi = \frac{dN_i}{\nu_i} \qquad (7.6)$$

where $\nu_i$ is a stoichiometric coefficient, negative for reactants and positive for products. The Gibbs energy variation with $\xi$ is plotted in Fig. 7.1. The reacton Gibbs energy, $\Delta_r G$, is obtained as the derivative of the mixture Gibbs energy with respect to the extent of reaction,

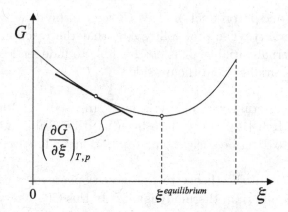

Figure 7.1: Gibbs energy of a reacting mixture as a function of the extent of reaction. The reaction Gibbs energy is the tangent to the curve.

illustrated for one composition by the tangent to the curve in Fig. 7.1.

$$\Delta_r G \equiv \left(\frac{\partial G}{\partial \xi}\right)_{T,p} = \sum_{i=1}^{n} \nu_i \mu_i \qquad (7.7)$$

This is the reaction Gibbs energy used in Eq. (3.3).

A reaction that spontaneously converts reactants to products has a negative slope. A positive slope means that the backward reaction occurs spontaneously. At equilibrium $\Delta_r G = 0$. This state is represented by the minimum of the curve in Fig. 7.1 at $\xi = \xi^{equilibrium}$.

We are concerned with reactions in incompressible liquid phases. Appendix A.3 describes how to calculate the chemical potential in general and in particular for chemical reactions. We use Eqs. (A.79) and (A.80) in this Appendix for the chemical potentials as a function of fugacities $f_i$ for the mixture with the components A, B, C and D, and obtain

$$\Delta_r G = \mu_C^{\ominus} + \mu_D^{\ominus} - \mu_A^{\ominus} - \mu_B^{\ominus} + RT \ln \frac{f_C f_D}{f_A f_B} \qquad (7.8)$$

where the standard pressure $p^{\ominus} = 1$ bar, and $\mu_i^{\ominus}$ is the chemical potential of component $i$ in the standard state with a pressure of 1 bar. At equilibrium, when $\Delta_r G = 0$,

$$\Delta_r G^{\ominus} = -RT \ln K$$

$$K = \left( \frac{f_C f_D}{f_A f_B} \right)_{eq} = \left( \frac{x_C x_D}{x_A x_B} \frac{\gamma_C \gamma_D}{\gamma_A \gamma_B} \right)_{eq} \qquad (7.9)$$

where $K$ is the dimensionless thermodynamic equilibrium constant and $x_i$ and $\gamma_i$ are the mole fraction and the activity coefficient of component $i$, respectively. The general form of Eq. (7.8) is

$$\Delta_r G = \sum_{i=1}^{n} \nu_i \mu_i^{\ominus} + RT \ln \left( \prod_{i}^{n} f_i^{\nu_i} \right) \qquad (7.10)$$

For the equilibrium constant this gives

$$K = \prod_{i}^{n} f_{i,eq}^{\nu_i} = \prod_{i}^{n} x_{i,eq}^{\nu_i} \gamma_{i,eq}^{\nu_i} \qquad (7.11)$$

The equilibrium constant, $K$, is used to replace $\Delta_r G^{\ominus}$ in Eq. (7.8). This gives

$$\Delta_r G = RT \ln \left( \prod_{i}^{n} \left( \frac{f_i}{f_{i,eq}} \right)^{\nu_i} \right) = RT \ln \left( \prod_{i}^{n} \left( \frac{x_i}{x_{i,eq}} \right)^{\nu_i} \left( \frac{\gamma_i}{\gamma_{i,eq}} \right)^{\nu_i} \right)$$
$$(7.12)$$

In an ideal mixture all $\gamma_i$ are equal to one.

The Gibbs-Helmholtz equation gives the temperature dependence of the equilibrium constant. For a reaction in a gas or liquid mixture at constant pressure

$$d \ln K = -\frac{\Delta_r H^{\ominus}}{R} d \left( \frac{1}{T} \right) \qquad (7.13)$$

The equilibrium constant can be obtained from pure component data. In order to calculate the driving force, we need data for the actual mixture.

**Exercise 7.1.1**

*Calculate the driving force for the esterification reaction at $T = 473\ K$ and $p = 2.5\ MPa$.*

$$\underbrace{CH_3COOH}_{\text{acetic acid}} + \underbrace{C_2H_5OH}_{\text{ethanol}} \rightleftharpoons \underbrace{CH_3COOC_2H_5}_{\text{ethyl acetate}} + H_2O$$

The composition is $x_{CH_3COOH} = 0.2$, $x_{C_2H_5OH} = 0.2$, $x_{CH_3COOC_2H_5} = 0.2$, and $x_{H_2O} = 0.4$, and the mixture is considered ideal. The Gibbs energies of formation at standard pressure and 298 K are given in the table

|  | $\Delta_f G^{\ominus}$ / kJ/mol | $\Delta_f H^{\ominus}$ / kJ/mol | $\nu_i$ | $x_i$ |
|---|---|---|---|---|
| acetic acid | −374 | −432 | −1 | 0.2 |
| ethanol | −168 | −235 | −1 | 0.2 |
| ethyl acetate | −328 | −445 | 1 | 0.2 |
| water | −228 | −242 | 1 | 0.4 |

- **Solution:** The standard reaction Gibbs energy calculated from the table is $(374 + 168 - 328 - 228) = -14$ kJ.mol$^{-1}$. The corresponding value for $\Delta_r G^{\ominus}/RT = -5.8$ so the equilibrium constant at 298 K becomes $K_{298} = 0.003$.

The standard reaction enthalpy according to the table is $\Delta_r H^{\circ} = (432 + 235 - 444 - 242) = -19.0$ kJ.mol$^{-1}$. We assume that the enthalpy of reaction is constant with temperature, and calculate $K$ for 473 K using the Gibbs-Helmholtz equation (7.13). This gives $\ln K_{473} = 5.8 - 2.8 = 3.0$. The exothermal reaction is shifted to the left by raising the temperature, in agreement with Le Chaterlier's principle. The driving force for the reaction at this temperature becomes

$$-\frac{\Delta_r G_{473}}{T} = R \ln K_{473} - R \ln \left( \prod_{i=1}^{n} (x_i)^{\nu_i} \right)$$

$$= 8.314 \text{ J.mol}^{-1}.\text{K}^{-1} \left( 3.0 - \ln \left( \frac{0.2 \times 0.4}{0.2 \times 0.2} \right) \right)$$

$$= 30.7 \text{ J.mol}^{-1}.\text{K}^{-1}$$

**Exercise 7.1.2** For the mixture specified in 7.1 calculate the equilibrium values of the mole fractions.

- **Solution:** For equilibrium conditions in the ideal mixture (all $\gamma_i = 1$), we have from Eq. (7.9)

$$K = \frac{x_C x_D}{x_A x_B}$$

Consider one mole of the initial mixture, $N^* = 1$ mole. The initial amount of any substance $N_i^*$ is then identical to the mole fraction, $N_i^* = x_i N^*$. At a given time, the extent of reaction is $\xi$. Any amount of substance is

$$N_i = N_i^* + \nu_i \xi$$

The total amount of substances is $N = \sum_{i=1}^{n} N_i = N^* + \xi \sum_{i=1}^{n} \nu_i$ and the mole fraction for any extent of reaction is

$$x_i = N_i^*/N + \nu_i \, \xi/N$$

In the considered case the amount of substance does not change, $N = N^* = 1$ mol. At equilibrium the extent of reaction is $\xi_{eq}$.

$$K = \frac{(0.1 + \xi_{eq} \text{ mol}^{-1})(0.7 + \xi_{eq} \text{ mol}^{-1})}{(0.1 - \xi_{eq} \text{ mol}^{-1})(0.1 - \xi_{eq} \text{ mol}^{-1})}$$

This quadratic equation in $\xi_{eq}$ can be solved using $K = 19.4$, and gives $\xi_{eq} = 0.030$ mol. The mole fractions are then $x_{C_2H_5OH} = 0.07$, $x_{CH_3COOH} = 0.07$, $x_{CH_3COOC_2H_5} = 0.13$, and $x_{H_2O} = 0.73$.

## 7.2 The reaction path

For a given reaction there is a continuous sequence of states between the pure reactant state with $\gamma = 0$ and the pure product state with $\gamma = 1$. The probability to find the reaction complex in the state $\gamma$ between these two states is given by $c(\gamma)$. The reacting complex changes its energy along the path. The left side of Fig. 7.2 shows an example of the energy of the reacting complex, $\Phi$, plotted versus $\gamma$. The picture shows that the reaction has an energy barrier, with a peak at a particular value of $\gamma$. The reaction must be activated to this level in order to proceed, even if the energy of the products are lower than that of the reactants. The fact that there exists an energy

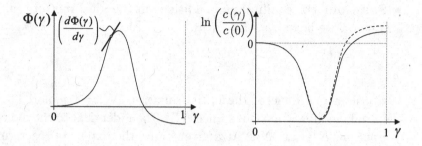

Figure 7.2: The two contributions to the chemical potential of the reaction mixture. The left diagram illustrates an energy barrier of a chemical reaction. The right diagram shows the probability density for having molecules in a state characterized by the coordinate $\gamma$ (solid line). The dashed line applies to chemical equilibrium.

barrier for conversion of reactants into products is not manifested in Fig. 7.1.

The coordinate $\gamma$ is a *mesoscopic* measure for the progress of a reaction in any macroscopic state with any composition given by $\xi$. The mesoscopic state of the system is given by the probability density $c(\gamma)$ of a reacting complex to be in the state $\gamma$. This probability density is taken as internal variable.

### 7.2.1   The chemical potential

The ln of the probability distribution for the reacting complex is illustrated in Fig. 7.2 (right). The probability of finding reacting complexes at the peak of the energy barrier in $\Phi(\gamma)$ (left) is low.

The chemical potential $\mu(\gamma)$ has an entropic part and what we call an enthalpic part $\Phi(\gamma)$

$$\mu(\gamma) = \mu(0) + RT \ln \frac{c(\gamma)}{c(0)} + \Phi(\gamma) \qquad (7.14)$$

where $\mu(0)$ is the reference chemical potential, *cf.* Exercise 7.5.1. The probability distribution of $c(\gamma)$ is here not normalized. Instead of such a normalization, we choose $c(0)$ such that $\mu(0) = RT \ln c(0)$

and $\Phi(0) = 0$. This implies that $\Phi(1) = \mu(1) - RT \ln c(1)$. The activation energy for the forward reaction is measured with respect to $\Phi(0) = 0$.

In equilibrium, the chemical potential is constant along the reaction coordinate, $\mu_{eq} = \mu_{eq}(\gamma) = \mu_{eq}(0)$. Using Eq. (7.14) it then follows that

$$c_{eq}(\gamma) = c_{eq}(0) \exp\left(-\frac{\Phi(\gamma)}{RT}\right) \tag{7.15}$$

Equation (7.15) is used to illustrate the dashed line in Fig. 7.2 (right). In equilibrium only a few molecules reach the top of the barrier $\Phi(\gamma)$, so the probability of this state $c_{eq}(\gamma)$ is low. Also away from equilibrium (solid line in Fig. 7.2) the probability of a reacting complex $c(\gamma)$ is small at the high energy barrier. The potential profile $\Phi(\gamma)$ is the same for an equilibrium and a non-equilibrium system for all values of the composition $\xi$.

The value of $\mu(\gamma)$ at the boundaries is known. For a reaction $2A \rightleftharpoons B$, the reactants' chemical potential is $\mu(\gamma = 0) = 2\mu_A$ and the product $\mu(\gamma = 1) = \mu_B$. More general, the chemical potential $\mu(0)$ is

$$
\mu(0) = \sum_i^{\{\text{reactants}\}} |\nu_i|\, \mu_i = \sum_i^{\{\text{reactants}\}} |\nu_i|\, \left[\mu_i^{\ominus} + RT \ln f_i\right]
$$

$$
= \sum_i^{\{\text{reactants}\}} |\nu_i|\, \left[\mu_i^{\ominus} + RT \ln (x_i \gamma_i)\right] \tag{7.16}
$$

Here the activity coefficients $\gamma_i$ are defined by $f_i = x_i p^{\ominus} \gamma_i$ which reduces to $f_i = x_i \gamma_i$ for the standard state pressure $p^{\ominus} = 1$ bar. The chemical potential of the products is

$$
\mu(1) = \sum_i^{\{\text{products}\}} \nu_i \left[\mu_i^{\ominus} + RT \ln (x_i \gamma_i)\right] \tag{7.17}
$$

Together with Eq. (7.14) these two relations determine $c(0)$ and $c(1)$. The standard state of $\mu(0)$ and $\mu(1)$ is the same as the standard states of the reactants together and the products together, respectively, see Eqs. (7.16) and (7.17).

## 7.2.2   The entropy production

The entropy production along the $\gamma$-coordinate is again the product of a flux $r(\gamma)$ and a driving force $-(1/T)(\partial\mu(\gamma)/\partial\gamma)$:

$$\sigma(\gamma) = -r(\gamma)\frac{1}{T}\frac{\partial\mu(\gamma)}{\partial\gamma} \tag{7.18}$$

This equation can be compared to Eq. (7.1). It has the same form, but is now written on a smaller scale, the mesoscale. The resulting flux-force relation is

$$r(\gamma) = -\frac{l(\gamma)}{T}\frac{\partial\mu(\gamma)}{\partial\gamma} \tag{7.19}$$

The variation in the chemical potential for a reaction complex is illustrated in Fig. 7.3. We see the reactant and product states for a reaction, say of $2A \rightleftharpoons B$, and the smooth transition between the states. The entropy production in $\gamma$-space is everywhere positive.

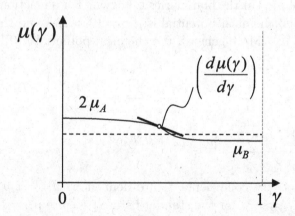

Figure 7.3: The variation in the chemical potential for a reaction ($2A \rightleftharpoons B$) along the mesoscopic reaction coordinate $\gamma$. The dashed line is for equilibrium.

## 7.3   A rate equation with a thermodynamic basis

The large (compared to $RT$) energy barrier in Fig. 7.2 hinders the chemical reaction. The conditions at the barrier peak are decisive for the overall rate. A quasi-stationary state thus develops, making the reaction rate constant along the coordinate, $r(\gamma) = r$. Using this property and the boundary conditions of Eqs. (7.16) and (7.17), we can integrate Eq. (7.18) to

$$\sigma_r = -r\frac{\mu(1) - \mu(0)}{T} = r(-\frac{\Delta_r G}{T}) \qquad (7.20)$$

which is equal to Eq. (7.1). The integrated entropy production along the $\gamma$-coordinate is equal to the macroscopic value, which it should be.

The chemical reaction can be seen as a diffusion process along the reaction coordinate over the activation energy barrier, see Fig. 7.2 (left). The conductivity $l(\gamma)$ in Eq. (7.19) is in good approximation proportional to the probability density $c(\gamma)$ [83], and we introduce a constant diffusion coefficient by

$$D_r \equiv \frac{Rl(\gamma)}{c(\gamma)} \qquad (7.21)$$

Substituting this relation into Eq. (7.19), using Eq. (7.14) and that $r$ is constant, we find

$$r = -\frac{D_r c(\gamma)}{RT}\frac{\partial\mu(\gamma)}{\partial\gamma}$$

$$= -\frac{D_r c(0)}{RT}\exp\left(-\frac{\Phi(\gamma)}{RT}\right)\exp\left(\frac{\mu(\gamma) - \mu(0)}{RT}\right)\frac{\partial\mu(\gamma)}{\partial\gamma}$$

$$= -D_r c(0)\exp\left(-\frac{\Phi(\gamma)}{RT}\right)\frac{\partial}{\partial\gamma}\exp\left(\frac{\mu(\gamma) - \mu(0)}{RT}\right) \qquad (7.22)$$

After multiplying this equation with $\exp\left(\Phi(\gamma)/RT\right)$, it can be integrated and we obtain

$$r = l_{rr}\left(1 - \exp\frac{\Delta_r G}{RT}\right) \qquad (7.23)$$

The flux-force relation becomes a non-linear one after integration over the internal coordinate. This is the preferred expression for the reaction rate to be used in the entropy production of Chapters 3, 6 and 11. The coefficient $l_{rr}$ is the macroscopic conductivity which is given by

$$
l_{rr} = \frac{1}{r_{rr}} = D_r c(0) \left[ \int_0^1 \exp \frac{\Phi(\gamma)}{RT} \mathrm{d}\gamma \right]^{-1}
$$

$$
= D_r \exp \frac{\mu(0)}{RT} \left[ \int_0^1 \exp \frac{\Phi(\gamma)}{RT} \mathrm{d}\gamma \right]^{-1}
$$

$$
= D_r \left[ \int_0^1 \exp \frac{\Phi(\gamma)}{RT} \mathrm{d}\gamma \right]^{-1} \prod_i^{\{\text{reactants}\}} \exp \frac{|\nu_i|\,\mu_i}{RT}
$$

$$
= D_r \left[ \int_0^1 \exp \frac{\Phi(\gamma)}{RT} \mathrm{d}\gamma \right]^{-1} \prod_i^{\{\text{reactants}\}} f_i^{|\nu_i|} \exp \frac{|\nu_i|\mu_i^{\ominus}}{RT} \qquad (7.24)
$$

Here we used Eq. (7.16). The coefficient $l_{rr}$ does not depend on $\gamma$.

We see that Eq. (7.23) gives $r = 0$ for the equilibrium condition $\Delta_r G = 0$, and that a Taylor expansion of the exponential gives the linear relation (7.3) close to equilibrium, when $\Delta_r G \ll RT$. The coefficient $l_{rr}$ can be determined from experiments.

## 7.4   The law of mass action

We shall see that the thermodynamic rate equation, Eq. (7.23), contains the law of mass action, which has been documented by numerous observations since its discovery in 1864 [29]. The law of mass action was first derived, using the mesoscopic description given in the previous Section, by Prigogine and Mazur for ideal mixtures, see [12, 82].

In the field of reaction kinetics [80], the rate has two contributions. The forward reaction rate is according to collision theory proportional to a kinetic coefficient and the concentrations of the reactants. The reverse reaction is proportional to a kinetic coefficient and the concentrations of the products.

By introducing Eqs. (7.16), (7.17) and the expression for $l_{rr}$ in the reaction rate, we obtain

$$r = D_r \left[ \int_0^1 \exp \frac{\Phi(\gamma)}{RT} d\gamma \right]^{-1}$$

$$\times \left\{ \prod_i^{\{\text{reactants}\}} f_i^{|\nu_i|} \exp \frac{|\nu_i|\mu_i^{\ominus}}{RT} - \prod_i^{\{\text{products}\}} f_i^{\nu_i} \exp \frac{\nu_i\mu_i^{\ominus}}{RT} \right\}$$

or

$$r = D_r \underbrace{\frac{\exp\left(\sum_i^{\{\text{reactants}\}} \frac{|\nu_i|\mu_i^{\ominus}}{RT}\right)}{\left[\int_0^1 \exp \frac{\Phi(\gamma)}{RT} d\gamma\right]}}_{k_f}$$

$$\times \left\{ \prod_i^{\{\text{reactants}\}} f_i^{|\nu_i|} - \underbrace{\exp\left(\sum_{i=1}^n \frac{\nu_i\mu_i^{\ominus}}{RT}\right)}_{1/K} \prod_i^{\{\text{products}\}} f_i^{\nu_i} \right\}$$

which gives

$$r = k_f \prod_i^{\{\text{reactants}\}} f_i^{|\nu_i|} - k_r \prod_i^{\{\text{products}\}} f_i^{\nu_i} \qquad (7.25)$$

This is the general form of the law of mass action. Values of the kinetic coefficients vary by many orders of magnitude for various systems, mainly due to different heights of the energy barrier. The energy barrier typically has a narrow maximum which is large compared to $RT$ at the so-called transition state $\gamma_0$. We may therefore write

$$\int_0^1 \exp \frac{\Phi(\gamma)}{RT} d\gamma = A \exp \frac{\Phi(\gamma_0)}{RT} \qquad (7.26)$$

This factor leads to the well known Arrhenius behavior of the kinetic coefficients.

The forward and reverse kinetic coefficients are related by

$$k_r = k_f/K \qquad (7.27)$$

This relation between the kinetic coefficients ensures that the reaction rate vanishes in the equilibrium state. Only one kinetic coefficient is independent in Eq. (7.25).

For ideal mixtures with $\gamma_i = 1$, Eq. (7.25) can be written in the form familiar from reaction kinetics

$$r = k_f \overset{\{\text{reactants}\}}{\underset{i}{\prod}} c_i^{|\nu_i|} - k_r \overset{\{\text{products}\}}{\underset{i}{\prod}} c_i^{\nu_i} \qquad (7.28)$$

This shows how the thermodynamic equation for $r$ can be reduced to the law of mass action for ideal mixtures, as first given by Guldberg and Waage [29].

## 7.5   The entropy production on the mesoscopic scale

We have postulated the local entropy production $\sigma(\gamma)$ as a product of a flux and the driving force along the $\gamma$-coordinate. We can substantiate this by analyzing the entropy production. We study the properties of the mixture at one point in Fig. 7.1 and along the coordinate. We assume local equilibrium for any mesoscopic state $\gamma$, so that the Gibbs equation is valid locally. The differential of the entropy due to a change $\delta c(\gamma)$ in the probability distribution is given by

$$\delta s = -\frac{1}{T} \int_0^1 \mu(\gamma)\delta c(\gamma)\mathrm{d}\gamma \qquad (7.29)$$

We did not consider changes in internal energy or volume in this expression, and we do not indicate the time dependence explicitly, to simplify notation. When the mixture is in chemical equilibrium,

$$\delta s = -\frac{1}{T} \int_0^1 \mu_{eq}(\gamma)\delta c(\gamma)\mathrm{d}\gamma = 0 \qquad (7.30)$$

Mass is conserved, so

$$\int_0^1 \delta c(\gamma)\mathrm{d}\gamma = 0 \qquad (7.31)$$

It follows from Eqs. (7.30) and (7.31) that the equilibrium chemical potential is independent of $\gamma$. It is constant, as illustrated in Fig. 7.3.

The entropy change which follows from Eq. (7.29) is

$$\frac{\partial s}{\partial t} = -\frac{1}{T}\int_0^1 \mu(\gamma)\frac{\partial c(\gamma)}{\partial t}\mathrm{d}\gamma \qquad (7.32)$$

The number of reacting complexes is constant, so the time rate of change of the probability can be written as

$$\frac{\partial c(\gamma)}{\partial t} = -\frac{\partial r(\gamma)}{\partial \gamma} \qquad (7.33)$$

By substituting this equation into Eq. (7.32) we obtain the entropy production integrated over the mesoscopic scale

$$\frac{\partial s}{\partial t} = \frac{1}{T}\int_0^1 \mu(\gamma)\frac{\partial r(\gamma)}{\partial \gamma}\mathrm{d}\gamma \qquad (7.34)$$

Partial integration then gives

$$\frac{\partial s}{\partial t} = \frac{1}{T}\int_0^1 \frac{\partial \mu(\gamma)r(\gamma)}{\partial \gamma}\mathrm{d}\gamma - \frac{1}{T}\int_0^1 r(\gamma)\frac{\partial \mu(\gamma)}{\partial \gamma}\mathrm{d}\gamma$$

$$= -\frac{\mu(0)}{T}r(0) + \frac{\mu(1)}{T}r(1)$$

$$- \frac{1}{T}\int_0^1 r(\gamma)\frac{\partial \mu(\gamma)}{\partial \gamma}\mathrm{d}\gamma \qquad (7.35)$$

The entropy flux along the $\gamma$-coordinate is given by

$$J_s(\gamma) = -\frac{\mu(\gamma)r(\gamma)}{T} \qquad (7.36)$$

We see in Eq. (7.35) that the integral of the divergence of this flux along the $\gamma$-coordinate gives the difference of the entropy flux into the $\gamma$-coordinate at $\gamma = 0$ minus the flux out of the coordinate at $\gamma = 1$. The second term of Eq. (7.35) gives the local entropy production along the $\gamma$-coordinate; it is Eq. (7.18).

**Exercise 7.5.1** *The potential $\Phi(\gamma)$ can in principle depend on the temperature and the pressure in the system. When $\Phi(\gamma)$ is assumed independent of both, show that the second and third terms in Eq. (7.14) are entropic and enthalpic contributions to the chemical potential, respectively.*

- **Solution:** When $\Phi(\gamma)$ is not a function of temperature and pressure, the entropy difference of a reaction complex in state $\gamma$ and state $\gamma = 0$ becomes

$$s(\gamma) - s(0) = -\frac{\partial(\mu(\gamma) - \mu(0))}{\partial T} = -R\ln\frac{c(\gamma)}{c(0)} \qquad (7.37)$$

  It follows that the enthalpy difference along the reaction coordinate is given by

$$h(\gamma) - h(0) = \mu(\gamma) - \mu(0) + T\left(s(\gamma) - s(0)\right)' = \Phi(\gamma) - \Phi(0) \qquad (7.38)$$

  The reaction enthalpy is therefore $\Delta_r H = \Phi(1) - \Phi(0) = \Phi(1)$ with the reference chosen.

## 7.6  Concluding remarks

We have shown that chemical reactions can be given a thermodynamic basis through mesoscopic non-equilibrium thermodynamics. The chemical reaction can then be dealt with in the same systematic way as we can deal with other transport processes. All transport laws can then be derived from the entropy production in the system. This is an advantage when we want to minimize the entropy production in process equipments, see Chapter 11.

In this derivation, the reacting mixture has been characterized by a probability distribution over the available states along the reaction coordinate $\gamma$. The reaction proceeds by diffusion of the reaction complex over an activation energy barrier. The rate is an explicit non-linear function of the driving force. When derived from this thermodynamic basis, the law of mass action contains fugacities rather than concentrations as variables.

# Chapter 8

# The lost work in the aluminum electrolysis

*The purpose of this chapter is to map the entropy production in an industrial process and to discuss the information in such a map. The lost work in a 230 kA aluminum electrolysis cell is estimated from the entropy production in the various parts of the cell, and the thermodynamic efficiency is calculated. The entropy production due to charge and to heat transfer are both large. The first can be changed by changing materials and electrode distances. The second can be changed by increasing cell dimensions.*

A good electrolysis cell in the 1980–90's used about 13 kWh of electric energy per kg of aluminum produced [84], down from a value between 14 and 18 kWh.$(kg_{Al})^{-1}$ 10 years earlier [85]. The minimum energy needed for the electrolysis at room temperature is only 5.4 kWh.$(kg_{Al})^{-1}$. The lost work or the exergy destruction footprint is the difference between these numbers. In this chapter we study the reasons for the lost work (the entropy production) and discuss possibilities for a reduction. Insight into this can be useful for cell design.

Good use of electric energy is central to the aluminum electrolysis industry, as electricity prices are central. An accurate control of the heat released by the cell is also crucial, in order to avoid cell disruption by crust melting and metal leakage through the cell wall. Much work has therefore been spent to determine the energy balance [86]. In this chapter we discuss also the entropy balance.

## 8.1   The aluminum electrolysis cell

Aluminum is produced from alumina, $Al_2O_3(s)$. Alumina is dissolved in molten cryolite, $Na_3AlF_6(l)$ before electrolysis. An oxygen-containing complex, probably $Al_2O_2F_6^{4-}$, reacts with the anode carbon (C) to give $CO_2$ (g), and to some degree also CO [87, 88]. The overall reaction is:

$$\frac{1}{6}Al_2O_3(s) + \frac{1}{4}C(s) \rightarrow \frac{1}{3}Al(l) + \frac{1}{4}CO_2(g) \qquad (8.1)$$

Since the bulk anode is in contact with air, there is also some excess carbon consumption:

$$C(s) + \frac{1}{2}O_2(g) \rightarrow CO(g) \qquad (8.2)$$

The $CO_2$ gas stirs the bulk melt as it escapes and is collected together with the CO gas at the top of the cell. The cell is illustrated in Fig. 8.1. Aluminum is formed at the bottom of the cell. The metal layer on top of the carbon block forms the bulk cathode. The product is collected at regular intervals, typically once a day. While the feeding of alumina is nearly continuous, the tapping of aluminum and the replenishing of the anode carbon, are not, so the cell does not operate under steady state conditions. An operation close to such conditions is an aim, however.

The electrolyte is contained in the central part of the cell, on top of the aluminum pool. The bulk anode consists of carbon blocks and steel connectors. Insulating material (alumina) covers the anode. The electrodes are prebaked (pre-made), and the anode is replaced as it is used. The cathode carbon conducts electric current from the metal to steel collector bars at the bottom of the cell. Many such cells are sequenced and the electric current goes from one cell via electric connectors to the next. The insulation at the bottom and on the sides consists of refractories, which are heat-resistant bricks. The cell potential, measured between the anode beams of two neighboring cells, is 4.1 V. The anode – cathode distance is 4.50 cm. The anode surface area is 30 m$^2$, while the cathode surface area is 50 m$^2$,

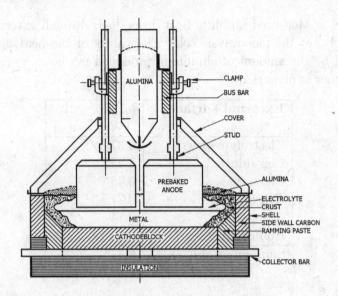

Figure 8.1: Cross section of the aluminum electrolysis cell. The metal is formed between the carbon anode and carbon bottom. Frozen electrolyte (crust) insulates the melt from the side walls. (Courtesy of K. Grjotheim).

giving anodic and cathodic current densities of $j = 7.7 \times 10^3$ and $4.6 \times 10^3$ A.m$^{-2}$, respectively.

The electrolyte has an average temperature of 960 °C and starts to freeze at 949 °C. A current efficiency of $y = 0.95$ is typical. The current efficiency is defined as the fraction of the electrons which reduce Al$^{3+}$. These data are typical of a 230 kA $(= I)$ cell of Hydro Aluminum in Øvre Årdal, Norway. The amount of aluminum produced per hour in such a cell is $m = yIM_{Al}/3F = 73.3$ kg.h$^{-1}$, where $M_{Al} = 27$ is the molar mass of aluminum in g.mol$^{-1}$. It is customary in the industry to refer all variables to $m$, and we shall also do this here. This will enable us to compare the performance of plants.

The frozen side ledge (crust) prevents contact between the molten corrosive melt and the side lining that consists of carbon, refractories and steel. It is imperative that the side ledge remains frozen for all conditions. This means that a careful control of the heat fluxes out of the cell is necessary. The heat fluxes through the cell surfaces have

Table 8.1: Measured absolute heat flows, $|q_k|$, through external surfaces, $A_k$, of the electrolysis cell. The values of the heat flows are divided by the amount of aluminum produced per hour, $m$ (see text for further explanation).

| External surface | $|q_k|/m$ kWh$^{-1}$.kg$_{Al}$] |
|---|---|
| Electrolyte cover | 0.37 |
| Cell sides | 2.42 |
| Cell ends | 0.48 |
| Cell bottom | 0.71 |
| Anode top | 2.14 |
| Iron conductors | 0.35 |
| Sum, $|q|/m$, Eq. (8.4) | 6.5 |

been measured. The heat flux through surface number $k$ is $J'_{q,k}$ (in kW.m$^{-2}$) and the surface area is $A_k$ (in m$^2$). The heat flow through one surface is then $q_k = J'_{q,k}A_k$, see Table 8.1, and the total heat flow from the cell per kg of aluminum is

$$q/m = \sum_k q_k/m = \sum_k J'_{q,k}A_k/m = 6.5 \text{ kWh.(kg}_{Al})^{-1} \qquad (8.3)$$

## 8.2   The thermodynamic efficiency

The thermodynamic efficiency gives an overall perspective on the energy transformation in the cell. Reactants enter and products leave the factory at temperature $T_0$ and pressure $p_0$; variables that define the environment (compare Section 2.3). Electric work, $w_{el}$, is used to accomplish the cell reaction. The reactants are in practice heated to the temperature of the bath, where the reaction takes place, and the products are subsequently cooled to the temperature of the environment. The change in internal energy between products and reactants at $T_0$ and $p_0$ is $\Delta U$. Some energy is also used to do mechanical work on the surroundings, $p_0\Delta V$. A large amount of heat, $q$, (*cf.* Table 8.1) is given off to the surroundings. The first law of thermodynamics gives:

$$\Delta U = q - p_0\Delta V + w_{el} \qquad (8.4)$$

where $q < 0$. The electric work added, per kg of aluminum is:

$$w_{el}/m = I\Delta\phi/m = 12.9 \text{ kWh.}(kg_{Al})^{-1} \tag{8.5}$$

where $I$ is again the electric current (230 kA) and $\Delta\phi$ is the measured cell potential (4.1 V).

The second law of thermodynamics distinguishes between the minimum work needed to do the process, and the work that is actually used. The difference is the lost work:

$$w_{el} - w_{el,min} = w_{lost} \tag{8.6}$$

The minimum work that is needed to perform the process at $T_0$ and $p_0$ is the reaction Gibbs energy. This quantity is found from the reversible cell potential at these conditions, $\Delta\phi_{rev} = -1.73$ V [86], giving $w_{el,min}/m = 5.4 \text{ kWh.}(kg_{Al})^{-1}$. The thermodynamic efficiency is therefore

$$\eta_{II} = \frac{w_{el} - w_{lost}}{w_{el}} = \frac{w_{el,min}}{w_{el}} = 0.42 \tag{8.7}$$

where the factor $m$ drops out. The second law of thermodynamics makes a distinction between energy forms; ranging their ability or potential to do work. Heat production at a high temperature is more valuable than heat production at the temperature of the surroundings, because the former energy can be used to do work. The difference between the first and the second laws efficiencies is their way of dealing with the reaction entropy. The entropy change is included in $w_{el,min} = -\Delta_1 G$ in Eq. (8.7). The subscript 1 refers to the main reaction.

The difference between the real work input and the minimum requirement divided by $m$, is 7.5 kWh.$(kg_{Al})^{-1}$. This is the total lost work in the electrolysis cell per kg produced (see Section 2.3):

$$w_{lost} = T_0 \frac{dS_{irr}}{dt} \tag{8.8}$$

In order to find the origin of the losses, or the exergy destruction footprint, *cf.* Section 2.4, we shall calculate $dS_{irr}/dt$ for all parts of the cell. A first step toward a possible reduction of losses is knowledge about where and how work is lost in the cell.

**Remark 9** *The first law of thermodynamics places all forms of energy changes on an equal footing; that is, all forms are equivalent in the energy balance. Heat leaving the electrolyte at $T_c = 960\ ^oC$ is then equivalent to heat given to the surroundings at the lower temperature $T_0 = 25\ ^oC$.*

The enthalpy, $\Delta_1 H$, as well as the entropy, $\Delta_1 S$, of reaction (8.1) are positive. The reaction Gibbs energy, $\Delta_1 G$, is accordingly smaller than $\Delta_1 H$. Part of the ohmic heat that is produced in the cell, is used to compensate the reversible heating need, equal to $T_c \Delta_1 S$. Heat produced in excess of this need is given off to the surroundings, leading to the values in in Table 8.1, where $|q|/m = 6.5$ kWh.$(\text{kg}_{\text{Al}})^{-1}$. For this work-consuming process, the first law efficiency is the ratio of energy gained by the process and the energy (work) that is actually used, compare Section 2.3. We obtain

$$\eta_I = \frac{\Delta U(T_0) + p_0 \Delta V}{w_{\text{el}}} = \frac{\Delta H(p_0, T_0)}{w_{\text{el}}} = 0.50 \qquad (8.9)$$

Some internal heat production compensates $T_c \Delta_1 S$, but we do not know where and how it is beneficial to reduce $|q|$. This knowledge is obtained from Eq. (8.8).

## 8.3   A simplified cell model

In order to make a first estimate of the cell's entropy production, we simplify the cell geometry. A cross-section of the cell in Fig. 8.1 is presented in Fig. 8.2. The entropy production for coupled transport of heat, mass and charge was given in Chapter 3. We assume constant fluxes perpendicular to cross-sectional areas $\Omega_j$ in the cell, and obtain for all bulk materials

$$\frac{dS_{\text{irr}}}{dt} = \iint \sigma d\Omega dx$$

$$= \sum_j q_j \Delta \left( \frac{1}{T_j} \right) + \sum_i J_i \Omega_i \left( -\frac{\Delta \mu_{i,T}}{T} \right) + I \left( -\frac{\Delta \phi_j}{T} \right) \qquad (8.10)$$

Here $q_j$ is the total heat flow through an area, and $I$ is the electric current. The conjugate forces are given by the parentheses. The

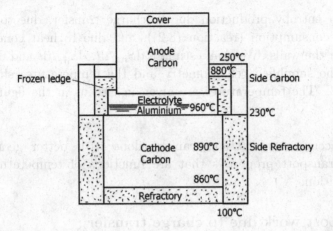

Figure 8.2: Schematic illustration of the cell showing typical temperatures.

entropy production of a reaction is:

$$\frac{dS_{irr}^{r}}{dt} = r\left(-\frac{\Delta G}{T_c}\right) V \tag{8.11}$$

The reaction rate $r$ of Eq. (8.2) is related to the number of faradays transferred through the cell. The value is calculated for the process volume.

In the electrode surfaces the entropy production is [89, 90]:

$$\frac{dS_{irr}^{s}}{dt} = q^{i}\Delta_{i,s}\left(\frac{1}{T}\right) + q^{o}\Delta_{s,o}\left(\frac{1}{T}\right) - I\frac{\Delta_{i,o}\phi}{T}$$

$$- \sum_{i} J_{i}^{i}\Omega_{i}\frac{\Delta_{i,s}\mu_{i,T}}{T} - \sum_{i} J_{i}^{o}\Omega_{o}\frac{\Delta_{s,o}\mu_{i,T}}{T} \tag{8.12}$$

The heat flow into the surface is $q^{i}$, and the heat flow out of the surface is $q^{o}$. The mass fluxes $J_{i}^{i}$ are directed into the surface, and $J_{i}^{o}$ are directed out of the surface. The potential difference $\Delta_{i,o}\phi$ is the surface potential drop. For details on the construction of the excess entropy production of the surface of these electrodes, see Hansen and Kjelstrup [89].

The total entropy production shall be obtained by adding results using these equations in the different parts of the cell. We distinguish between entropy production due to charge transfer, due to excess carbon consumption (reaction (8.2)), and due to heat conduction through the walls. We shall estimate $dS_{irr}/dt$, $dS_{irr}^s/dt$ and $dS_{irr}^r/dt$ using the simplified cell geometry and the temperatures shown in Fig. 8.2. The temperatures are given in Celsius in the figures and tables.

More accurate calculations can be done with better geometries using transport properties that are functions of temperature and composition.

## 8.4   Lost work due to charge transfer

Consider first the losses connected with the electric circuit. We shall see below that these can be estimated to 2.4 kWh.$(kg_{Al})^{-1}$. The losses in the iron bars that connect the series of single cells, are small in this context. From this point out, we will replace the symbol $w$ with $w'$ to indicate that the special dimension, kWh.$(kg_{Al})^{-1}$, is used.

### 8.4.1   The bulk electrolyte

There are losses related to the potential drop across the electrolyte. These losses are mainly ohmic. There is a good stirring of the electrolyte from $CO_2(g)$ which escapes the electrolyte, and from the circular movement of the metal in the very strong magnetic field caused by the electric current. An electric potential drop of $-1.7$ V is normal, where $-1.5$ V is due to the main part of the electrolyte and $-0.2$ V is assumed for the bulk part of the melt that contains gas bubbles. The last term of Eq. (8.10) gives

$$w'_{lost,1} = \frac{T_0}{T_c m} I(-\Delta\phi_1) = 1.30 \text{ kWh.}(kg_{Al})^{-1} \qquad (8.13)$$

### 8.4.2   The diffusion layer at the cathode

The main charge carrier in the melt is $Na^+$. Close to the cathode, $Na^+$ accumulates, and diffuses back. The diffusion layer with gradients in chemical potential and in electric potential, is approximately 1 mm thick $(= \Delta x)$. We can neglect temperature gradients in this layer. With the electrode surface as a frame of reference for transport,

3 $Na^+$ ions move away from the surface in exchange for one $Al^{3+}$ ion moving to the surface. The steady state flux of $AlF_3$ into the surface is therefore $J_{AlF_3} = j/3F$, while the net flux of NaF is zero. The entropy production for the layer is

$$\left(\frac{dS_{irr}}{dt}\right)_2 = -\frac{I}{T_c}\left(\frac{\Delta\mu_{AlF_3}}{3F} + \Delta\phi_2\right) \tag{8.14}$$

The current density at stationary state is therefore

$$\frac{I}{\Omega} = -\frac{\kappa}{\Delta x}\left(\frac{\Delta\mu_{AlF_3}}{3F} + \Delta\phi_2\right) \tag{8.15}$$

where $\kappa$ is the conductivity of the stationary state boundary layer, 19 kohm.m$^{-1}$ [91]. The lost work is

$$w'_{lost,2} = \frac{T_0}{T_c m\kappa}\frac{I^2}{\Omega}\Delta x = 0.05 \text{ kWh.}(kg_{Al})^{-1} \tag{8.16}$$

### 8.4.3 The electrode surfaces

The anode overpotential gives the major lost work at the electrode surfaces. A typical overpotential, $\eta$, is 0.50 V [86]. Hansen [89, 90] estimated thermal and chemical forces as well as transport coefficients consistent with this value. Small forces ($< 5$ K$^{-1}$, and $< 1$ J.K$^{-1}$.mol$^{-1}$) were obtained. In the stationary state, the mass fluxes relative to the fluoride ion frame of reference (or the surface frame of reference) is zero. The lost work from Eq. (8.12) was therefore essentially given by the overpotential, which amounts to

$$w'_{lost,s} = 0.48 \text{ kWh.}(kg_{Al})^{-1} \tag{8.17}$$

The entropy production at the cathode was negligible [89, 90], consistent with a small overpotential at this electrode.

### 8.4.4 The bulk part of the anode and cathode

The carbon parts of the anode and cathode conduct heat and charge. From Eq. (8.10) we obtain (see Refs. [92, 93] for more details):

$$\left(\frac{dS_{irr}}{dt}\right)_j = q_j\Delta\left(\frac{1}{T_j}\right) + \frac{I}{T}(-\Delta\phi_j) \tag{8.18}$$

Table 8.2: The lost work in the anode and cathode at average temperature 1200 K.

|  | Anode | Cathode |
|---|---|---|
| $\lambda$ W$^{-1}$.K.m | 10 | 13 |
| $T_h - T_l$ K$^{-1}$ | 960-785 | 960-860 |
| $\Delta x_j$ m$^{-1}$ | 0.35 | 0.44 |
| $\Omega_j$ m$^{-2}$ | 30 | 50 |
| $\kappa$ ohm.m | 19 000 | 40000 |
| $\pi$ J$^{-1}$.K.mol | 1520 | 2446 |
| $w'_{\text{lost}}$ kWh$^{-1}$.kg$_{\text{Al}}$ | 0.2 | 0.08 |

The constitutive relations are

$$q_j = -\lambda_j \Omega_j \frac{\Delta T_j}{\Delta x_j} + \pi_j \frac{I}{F} \tag{8.19}$$

$$\Delta \phi_j = -\frac{\pi_j}{FT} \Delta T - \frac{\Delta x_j}{\Omega_j \kappa_j} I \tag{8.20}$$

When these equations are introduced into Eq. (8.18), we obtain for both electrodes:

$$\left( \frac{dS_{\text{irr}}}{dt} \right)_j = -\lambda_j \Omega_j \frac{\Delta T_j}{\Delta x_j} \Delta \left( \frac{1}{T_j} \right) + \frac{I^2}{\Omega_j} \frac{\Delta x_j}{\kappa_j T} \tag{8.21}$$

The input data for each electrode and the results of these calculations are given in Table 8.2. The thermal conductivities have been measured [86]. Heat is transferred in the bulk anode from 960 °C to an estimated 785 °C, and in the bulk cathode from 960 °C to an estimated 860 °C. The table shows that the lost work in the two electrodes are 0.20 and 0.08 kWh.(kg$_{\text{Al}}$)$^{-1}$, for the anode and cathode, respectively.

## 8.5   Lost work by excess carbon consumption

By excess carbon consumption, we mean carbon that disappears from the cell, in excess of what is used by reaction Eq. (8.1). The carbon disappears as CO by the reaction Eq. (8.2). The rate of the consumption of carbon was estimated from excess weight loss of anode material. The minimum amount of carbon needed for production of one ton of aluminum is 350 kg. The real consumption is typically 400 kg. This gives the rate of excess carbon consumption as 14% of the rate of reaction Eq. (8.1). The Gibbs energy change of reaction Eq. (8.2) is $-219.5$ kJ per mol of CO produced [86]. This gives

$$w'_{\text{lost,r}} = 0.10 \text{ kWh.}(\text{kg}_{\text{Al}})^{-1} \tag{8.22}$$

The work lost by the excess carbon consumption is small compared to other losses.

## 8.6   Lost work due to heat transport through the walls

The work that is lost by heat conduction through the walls of the container, can be calculated from the two temperatures that bound the walls and the total heat flow, $q/m = 6.5 \text{ kWh.}(\text{kg}_{\text{Al}})^{-1}$, see Table 8.1. Grossly speaking, we can use $T_0 = 25$ °C (the temperature in the hall) and $T_c = 860$ °C (see Fig. 8.2). This gives:

$$w'_{\text{lost}} = T_0 \frac{q}{m} \left( \frac{1}{T_0} - \frac{1}{T_c} \right) = 4.80 \text{ kWh.}(\text{kg}_{\text{Al}})^{-1} \tag{8.23}$$

This calculation of the lost work is not specific. It does not address the different modes of heat transfer. Heat is transferred by conduction across the cell walls, and by convection from the cell surface to the room. There is also black body radiation from the cell surface. We would like to know more about the specific contributions, and proceed to estimate them.

### 8.6.1   Conduction across the walls

Wall number $j$ has a thickness $\Delta x_{wj}$ and a cross-sectional area for heat transfer equal to $\Omega_{wj}$. The temperature difference across $\Delta x_w$ is $\Delta T_j$, and the thermal conductivity is $\lambda_j$. The entropy production in the wall materials is then

$$\left( \frac{dS_{\text{irr}}}{dt} \right)_j = -\lambda_j \frac{\Delta T_j}{\Delta x_{wj}} \Omega_{wj} \Delta \left( \frac{1}{T_j} \right) \tag{8.24}$$

Material data, temperatures and geometries are given in Table 8.2 and in Fig. 8.2. The temperature of the solid material on the electrolyte side, is the temperature of the electrolyte, 960 °C. The surface temperature of the container side is 230 °C, of the top crust it is

Table 8.3: Contributions to lost work from the sides of the container walls in Fig. 8.2. Symbol C means that the walls are lined with carbon, while R means refractory lining.

| | Top layer | Side R | Side C |
|---|---|---|---|
| $\lambda$ W$^{-1}$.K.m | 0.34 | 3.1 | 5.0 |
| $T_h - T_l$ K$^{-1}$ | 880-250 | 865-190 | 450-250 |
| $\Delta x_w$ m$^{-1}$ | 0.12 | 0.42 | 0.15 |
| $\Omega_w$ m$^{-2}$ | 15 | 18 | 13 |
| $w'_{lost}$ kWh$^{-1}$.kg$_{Al}$ | 0.11 | 0.46 | 0.19 |
| | **End R** | **End C** | **Bottom** |
| $\lambda$ W$^{-1}$.K.m | 0.76 | 4.4 | 0.4 |
| $T_h - T_l$ K$^{-1}$ | 880-110 | 470-255 | 860-90 |
| $\Delta x_w$ m$^{-1}$ | 0.35 | 0.15 | 0.29 |
| $\Omega_w$ m$^{-2}$ | 6 | 4 | 50 |
| $w'_{lost}$ kWh$^{-1}$.kg$_{Al}$ | 0.07 | 0.05 | 0.40 |

250 °C, and at the bottom of the pot, it is 100 °C. Between the side ledge and the side wall it is 450 °C. Below the cathode carbon it is 860 °C.[1] The temperature $T_h$ is the temperature of the surface facing the electrolyte, while $T_l$ is the temperature of the surface facing the outside of the container. Surface temperatures are given in Table 8.3. The values of the high and low temperature in each layer are given in Celsius.

The sum of lost work from the container walls (sum of the bottom line of Table 8.3) is $w'_{lost,4} = 1.4$ kWh.(kg$_{Al}$)$^{-1}$. There is a difference in lost work between the sides and the ends of the cell. This reflects, on one hand, the difference in area of these surfaces (the area ratio is 1:3), and on the other hand, the fact that relatively more entropy production takes place at the sides, because the current collector bars are located there. The lost work in the frozen side ledges was estimated by Hansen [90] to be 0.2 kWh.(kg$_{Al}$)$^{-1}$. These losses together account for 1.6 kWh.(kg$_{Al}$)$^{-1}$.

---

[1]The isotherms, from which these average temperatures were taken, were all generated for the accurate geometry in a numerical model of Hydro Aluminum ASA.

### 8.6.2 Surface radiation and convection

The sum of all calculated losses in the walls so far do not explain a total loss of 4.8 kWh.$(\text{kg}_{Al})^{-1}$. The outer walls of the container are cooled by radiation, convection and conduction. The energy flux by black body radiation from this is

$$J_q = \frac{cu}{4\pi} \qquad (8.25)$$

where $c$ is the velocity of light and $u$ is the energy density $u = \beta T^4$. With $\beta = 7.56 \times 10^{-16}$ J.m$^{-3}$.K$^{-4}$, $c = 3 \times 10^8$ m.s$^{-1}$, and using an estimated surface temperature of 390 K, we obtain $J_q = 23.4$ W.m$^{-2}$.

The entropy production due to radiation is obtained by multiplying this energy flux with the thermal driving force and the surface area (about 160 m$^2$). In this case there is no linear relation between the flux and the driving force, but the expression for the entropy production is the same, *cf.* Chapter 7.

The lost work by radiation is accordingly 67 kW or 0.9 kWh.$(\text{kg}_{Al})^{-1}$ for the present example. The estimate is uncertain because of the $T^4$-dependence. A 10% increase of the temperature of the surface gives about a 50% increase in the entropy production. It is also difficult to estimate the air circulation around the cell which takes heat away by convection.

Anyway, the lost work by radiation from the cell and convection between the wall surface and the surroundings, can be significant, and may explain the value obtained in Eq. (8.23).

### 8.7 The exergy destruction footprint

The lost work in all the different parts of the electrolysis cell are now known. Together they give the exergy destruction footprint of the process, *cf.* Section 2.4. The various parts are summarized in Table 8.4. Lost work from heat transport through the walls is taken as the value calculated in Section 8.6. We see then that the total loss, 7.4 kWh.$(\text{kg}_{Al})^{-1}$, is close to the difference between the real work and the minimum work $(12.9\text{-}5.4)$ kWh.$(\text{kg}_{Al})^{-1} = 7.5$ kWh.$(\text{kg}_{Al})^{-1}$. A

Table 8.4: **The exergy destruction footprint of the aluminum electrolysis cell.** All contributions are given in $kWh.(kg_{Al})^{-1}$. The sum is the accumulated loss.

| Loss type | Loss location | Amount lost | Sum |
|---|---|---|---|
| **Charge transfer** | electrolyte resistance | 1.3 | |
| | diffusional layers | 0.1 | |
| | electrode surfaces | 0.5 | |
| | bulk cathode | 0.1 | |
| | bulk anode | 0.2 | 2.2 |
| **Hot reactants** | Al and $CO_2$ | 0.3 | 2.8 |
| **Reaction 8.2** | anode | 0.1 | 2.9 |
| **Thermal** | walls, surroundings | 4.8 | 7.4 |

simplified geometry was, however, chosen for the model, so uncertainties are large.

The table gives a map of where and why the losses occur, and this can point to directions for further efforts to reduce or justify the losses.

The accumulated lost work in the charge conducting pathways is $2.5 \ kWh.(kg_{Al})^{-1}$. The ohmic losses in the bath gives the largest contribution to this type of lost work. The other major loss occurs at the anode surface. Around 72% of the lost work in the charge conducting pathways can be explained by these process steps.

Efforts that lower the electrolyte resistance have therefore been many. The ohmic losses in the electrolyte could be reduced by reducing the distance between the anode and the surface of the aluminum further. However, in practice this is very difficult due to the large magnetic fields that rotate the metal. The substantial entropy production in the anode surface has also made the anode a target of further attention. The electrode mechanism for the reaction is still largely unknown. The lost work connected to the electrode reaction cannot be avoided, but it can be changed. Inert anodes, that liberate oxygen in stead of carbon dioxide, have not yet been made.

The energy content in the hot products $(0.3 \text{ kWh.}(\text{kg}_{Al})^{-1})$ can be regained by heat exchange. The excess loss of carbon to the air in the electrolysis hall, *cf.* Eq. (8.2), is difficult to reduce.

The thermal losses by conduction and convection are by far the most substantial losses. The losses at low temperature dominate, and they depend largely on the location. For instance, without an extra alumina layer on top of the crust, there is an extra energy need of $0.3 \text{ kWh.}(\text{kg}_{Al})^{-1}$ to maintain the electrolyte temperature. But an (excess) heat leak at the top, reduces the heat flux through the cell sides. As a consequence, the side ledge crusts may grow in thickness, and lead to more lost work.

The relatively large entropy production in the carbon parts of the electrodes can also be reduced, but this may also increase losses in other parts. A good cell design therefore means that the heat fluxes are distributed in a manner that gives minimum total lost work (see Chapter 11). The requirement that there is a side ledge of frozen melt to prevent the molten electrolyte from attacking the outer steel walls, limits the possibilities for a reduction.

## 8.8 Concluding remarks

We have seen above how the exergy destruction footprint of a process can point to directions for further work, or even different ways to operate the process with present-day materials.

The losses listed in Table 8.4 are of two types. One type, i.e. losses connected with charge conduction, is (nearly) proportional to the electric current. The second are thermal type losses, which are not proportional to the electric current. These losses are mainly given by the heat flux and the temperature difference between the bath and the surroundings.

So how can this knowledge be used to improve the design? The electric current must run and the temperature difference between the bath and the surroundings cannot be changed. But the total heat transferred to the surroundings will depend on the surface area of the container. Here lies a possibility for the reduction of lost work per kg Al produced. The lost work in the charge conducting circuit is largely proportional to the amount of aluminum produced, but the lost work in the container walls is not. By making the cells larger, the lost work from heat transport in the walls can be reduced per kg Al produced. An argument in favor of larger production cells, see Grjotheim and Kvande [93], is therefore also an argument about a smaller exergy destruction footprint.

Since so much of the lost work can be related to heat transfer from the cell walls to the surroundings, methods to recover this waste heat may be beneficial, *cf.* Section 4.3. Large low-temperature thermal losses are typical in many industries. In the future, thermoelectric cells may become useful [61].

# Chapter 9

# Coupled transport through surfaces

*We derive the excess entropy production for transport of heat, mass and charge through an interface. Finite differences of intensive variables into, out of, and across the surface, as well as Gibbs reaction energies give the driving forces. Equivalent sets of flux-force pairs are derived. The flux-force relations are used to describe phase transitions and electrode phenomena.*

We are seeking a systematic way to set up equations of transport for heat, mass and charge into, out of, and across an interface or surface. Such equations will allow us to describe, for instance, electrochemical reactions or phase transitions of wide interest for industrial applications and for modeling of natural phenomena.

In homogeneous phases, thermodynamic variables are continuous. In contrast, at a surface, they may jump, and the fluxes and forces may become discontinuous. The reason for this is that a surface may act as a source or sink of energy of various sorts; thermal, electrical or chemical. This situation is typical for phase transitions, distillation, heterogeneous catalysis, or electrochemical reactions, to mention a few examples [94, 95, 96, 97, 31]. This chapter presents a way to deal with surface phenomena in terms of non-equilibrium thermodynamics. The method builds on Gibbs definition of a surface and the assumption of Bedeaux, Albano and Mazur [98, 99] that the

interface is a separate thermodynamic system. Thermodynamic relations between surface variables also remain valid when the system
is out of equilibrium. This means that we assume that the surface is
in local equilibrium.

Typical for an interfacial region is that its thickness is small compared to the thickness of the adjacent homogeneous phases. Thermodynamic properties of the surface can nevertheless be well defined
using Gibbs' *surface excess densities* of mass, entropy and energy [44].
The surface is then regarded as a two-dimensional, autonomous thermodynamic system. Following Gibbs' systematic procedure we will
derive the entropy production of the surface in Sections 9.1–9.3, and
proceed to apply the expression to a simple phase transition (Section
9.4) and to electrode reactions (Section 9.5).

More details on transport through surfaces were given by Kjelstrup
and Bedeaux [31, 100, 101]. As they have done, we shall also not consider transport along the surface ([98, 102, 99]). From the perspective
of non-equilibrium thermodynamics, surfaces are little explored.

## 9.1   The Gibbs surface in local equilibrium

Gibbs defined the *dividing surface* as "a geometrical plane, going
through points in the interfacial region, similarly situated with respect to conditions of adjacent matter". Many different positions
can be chosen for a plane of this type, but a position somewhere
between co-ordinates a and b in Fig. 9.1 is practical. The concentration of component A, $c_A$, varies along the $x$-axis and the *excess mass
density* of A is $\Gamma_A$

$$\Gamma_A(y, z)$$
$$= \int_a^b \left[ c_A(x, y, z) - c_A^g(a, y, z)\theta(d-x) - c_A^l(b, y, z)\theta(x-d) \right] dx \quad (9.1)$$

Here d is the position of the dividing surface. The excess density
is the *adsorption* (in $mol.m^{-2}$). It is in general a function of the
position $(y, z)$ along the surface. The Heaviside function, $\theta$, is by
definition unity when the argument is positive and zero when the
argument is negative. On a macroscopic scale the surface appears to

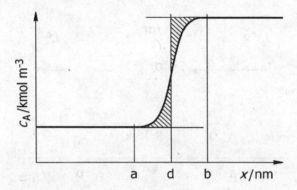

Figure 9.1: Determination of the position of the equimolar surface of component A. The vertical line is drawn so that the areas between the curve and the bulk densities are the same.

be two-dimensional. Nevertheless the surface may possess a temperature, chemical potential and other thermodynamic variables of its own.

All excess properties of a surface can be given by integrals like Eq. (9.1). The excess variables are the extensive variables of the surface. They describe the surface and how it *differs* from the adjacent homogeneous phases. One may shift the positions a and b further into the bulk phases without changing the adsorption. In this sense the precise locations of a and b are not important for the value of the adsorption. Certain experimental quantities, like the surface tension of a flat surface, do not depend on the choice.

The excess densities depend on the location of the dividing surface. The *equimolar surface* of component A is a special dividing surface, for which d is chosen such that the surplus of moles of the component on one side of the surface is equal to the deficiency of moles of the component on the other side of the dividing surface, *cf.* Fig. 9.1. This makes $\Gamma_A = 0$ for this surface. One can choose an equimolar dividing surface of any component. Other choices are also possible [44, 31, 103].

The assumption of local equilibrium [98, 99, 104] means that thermodynamic relations between surface excess densities remain valid when

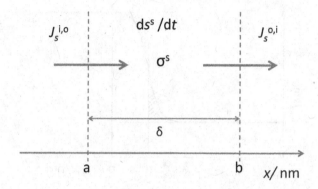

Figure 9.2: The change in the excess surface entropy density is due to entropy fluxes in and out of the surface and entropy production in the surface. The surface thickness, $\delta = b-a$, is small compared to the thickness of the adjacent phases.

the system is out of equilibrium. Away from global equilibrium, the surface constructed in this manner may have a different temperature and chemical potentials from the values in the adjacent homogeneous phases.

## 9.2   Balance equations

We are concerned with one-dimensional transport problems, and consider transport normal to the surface, in the $x$-direction. The second law applies to any volume element of a homogeneous phase (*cf.* Chapter 3). This statement is now extended to surface area elements. The entropy balance is

$$\frac{d}{dt}s^s(t) - J_s^{i,o}(t) + J_s^{o,i}(t) = \sigma^s(t) \geq 0 \qquad (9.2)$$

Here $ds^s(t)/dt$ is the rate of change of the excess entropy density per unit of surface area. This is equal to the entropy flux into the surface, minus the entropy flux out of the surface, plus the excess entropy production, $\sigma^s(t)$. The (asymptotic) value of the entropy flux in phase i, $J_s^{i,o}(t) \equiv J_s^i(a,t)$, is directed into the surface, while the entropy flux in phase o, $J_s^{o,i}(t) \equiv J_s^o(b,t)$, is directed out of the surface. The entropy balance is illustrated in Fig. 9.2. The notation

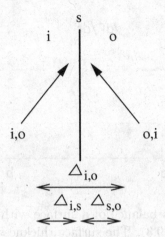

Figure 9.3: Notation used for variables that describe transport across surfaces.

is further illustrated in Fig. 9.3. According to the second law of thermodynamics, the excess entropy production is positive for the surface. We restrict ourselves to the case when the variables and fluxes depend only on $x, t$. The fluxes are all in the $x$-direction. As a consequence, the excess densities depend only on $t$ and we use straight d's when we differentiate.

We shall find explicit expressions for $\sigma^s(t)$, following the same procedure as for $\sigma(x, t)$ in the homogeneous phase (*cf.* Chapter 3). We shall combine mass balances and the energy balance with the local form of the Gibbs equation, and see that $\sigma^s(t)$ can also be written as the product sum of thermodynamic fluxes and forces. These are the *conjugate* fluxes and forces of the surface. Such sets of fluxes and forces can be used to describe transports across and into the surface. *The positive direction of transport is from smaller to larger $x$* (from left to right in the figures). We recommend the use of the symbol list for a check of the dimensions in the equations.

We use straight derivatives in Eq. (9.2), because the excess surface entropy density depends only on the time, not on the position. The first roman superscript gives the phase; i, s or o in this case. The second superscript, o or i, indicates the phase across the surface.

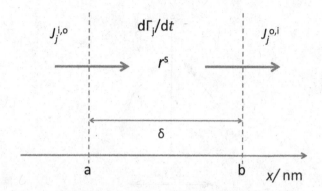

Figure 9.4: The mass balance of a surface with a chemical reaction $r^s$, according to Eq. (9.3). The surface thickness is $\delta =$ b$-$a.

The combination i,o means therefore the value in phase i as close as possible to the o-phase at the interface. The notation is illustrated in Fig. 9.3. Balance equations of the form Eq. (9.2) can be derived from the continuous description [105, 106]. We use the surface frame of reference in which the dividing surface is not moving.

The adsorption $\Gamma_j$ of component $j$ increases, when the influx of $j$ is larger than the out-flux or when a chemical reaction leads to a product:

$$\frac{\mathrm{d}}{\mathrm{d}t}\Gamma_j(t) = J_j^{i,o}(t) - J_j^{o,i}(t) + \nu_j r^s(t) \qquad (9.3)$$

The asymptotic value of the flux of $j$ from phase i into the surface is $J_j^{i,o}(t)$, while the asymptotic value of the flux out of the surface into phase o is $J_j^{o,i}(t)$, *cf.* Fig. 9.4. The excess chemical reaction rate in the surface element is $r^s(t)$, and $\nu_j$ is a stoichiometric constant. The surface is again the frame of reference for the fluxes. A flux is not necessarily present on both sides of the surface, meaning that some mass fluxes in Eq. (9.3) are zero. This is, for instance, the case for electrode surfaces.

At electrode surfaces, there is always a discontinuity in the charge carrier; typically from electron to ion or vice versa. The electrode reaction describes this change. The reaction Gibbs energy is

$$\Delta_r G^s(t) \equiv \sum_j \nu_j \mu_j^s \qquad (9.4)$$

The sum in Eq. (9.4) is over all species, charged or uncharged, involved in the reaction. In our description of the surface we shall also need the contribution to the reaction Gibbs energy due to the neutral species

$$\Delta_n G^s(t) \equiv \sum_{j \,\in\, neutral} \nu_j \mu_j^s \qquad (9.5)$$

where the sum in Eq. (9.5) is only over the neutral components. The difference between $\Delta_n G^s$ and the usual expression $\Delta_r G^s$, is due to the chemical potentials of ions and electrons. The surface may be polarized and include a double layer. Including the double layer, the surface is still electroneutral. The thermodynamic state of the surface is determined using absorptions of neutral components only. Electroneutrality in the homogeneous phases and of the surface means that the electric current density is independent of the position:

$$j^i(x,t) = j^o(x,t) = j(t) \qquad (9.6)$$

## Exercise 9.2.1

*Derive Eq. (9.3) using Fig. 9.4*

- **Solution:** In Fig. 9.4 we see that the change in the number of moles of a component is due to the flux in of this component at position a and out at position b of the surface area element. This gives

$$\frac{\mathrm{d}}{\mathrm{d}t} N_j^s(t) = -\Omega[J_j^o(b,t) - J_j^i(a,t)] \pm \Omega(b-a)\nu_j r(t)$$

where $\Omega$ is the surface area, normal to the flux direction. The fluxes are per unit of surface area and $r$ is the reaction rate per unit of volume between a and b. By dividing this equation left and right by the (time independent) surface area, defining $r^s(t) =$(b-a)$r(t)$ and identifying $J_j^i(a,t) = J_j^{i,o}(t)$ and $J_j^o(b,t) = J_j^{o,i}(t)$, we obtain Eq. (9.3).

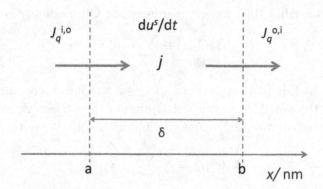

Figure 9.5: The surface energy balance. The change in the surface excess energy density is the difference in the total heat fluxes, and the electric work per unit of time and area done to the surface, *cf.* Eq. (9.7).

The first law of thermodynamics for the electroneutral surface is:

$$\frac{du^{s}(t)}{dt} = J_q^{i,o}(t) - J_q^{o,i}(t) - j(t)[\phi^{o,i}(t) - \phi^{i,o}(t)] \qquad (9.7)$$

The time rate of change in the excess energy density of the surface, $u^s$, is given by the asymptotic value of the total heat flux from the adjacent phase i into the surface from the left, $J_q^{i,o}$, minus the asymptotic value of the total heat flux out of the surface into the adjacent phase o, $J_q^{o,i}$, and the electric work per unit of time and area done to the surface, $-j(\phi^{o,i} - \phi^{i,o})$. Here $\phi^{o,i}$ and $\phi^{i,o}$ are the electric potentials in the homogeneous phases close to the surface. The change in energy density in the surface is illustrated in Fig. 9.5.

## 9.3   The excess entropy production

The Gibbs equation for the surface is

$$du^{s} = T^{s}ds^{s} + \sum_{j=1}^{n} \mu_{j}^{s}d\Gamma_{j} \qquad (9.8)$$

The densities are per unit area. From here on, the time dependence is not explicitly indicated. The time derivative of the excess entropy density is:

$$\frac{\mathrm{d}s^{\mathrm{s}}}{\mathrm{d}t} = \frac{1}{T^{\mathrm{s}}}\frac{\mathrm{d}u^{\mathrm{s}}}{\mathrm{d}t} - \frac{1}{T^{\mathrm{s}}}\sum_{j=1}^{n}\mu_j^{\mathrm{s}}\frac{\mathrm{d}\Gamma_j}{\mathrm{d}t} \qquad (9.9)$$

By introducing Eqs. (9.7) and (9.4) into Eq. (9.9), and comparing the result to the entropy balance, Eq. (9.2), we find the excess entropy production:

$$
\begin{aligned}
\sigma^{\mathrm{s}} &= J_q^{\mathrm{i,o}}\left(\frac{1}{T^{\mathrm{s}}} - \frac{1}{T^{\mathrm{i,o}}}\right) + J_q^{\mathrm{o,i}}\left(\frac{1}{T^{\mathrm{o,i}}} - \frac{1}{T^{\mathrm{s}}}\right) \\
&\quad + \sum_{j=1}^{n} J_j^{\mathrm{i,o}}\left[-\left(\frac{\mu_j^{\mathrm{s}}}{T^{\mathrm{s}}} - \frac{\mu_j^{\mathrm{i,o}}}{T^{\mathrm{i,o}}}\right)\right] \\
&\quad + \sum_{j=1}^{n} J_j^{\mathrm{o,i}}\left[-\left(\frac{\mu_j^{\mathrm{o,i}}}{T^{\mathrm{o,i}}} - \frac{\mu_j^{\mathrm{s}}}{T^{\mathrm{s}}}\right)\right] + j\left[-\frac{1}{T^{\mathrm{s}}}(\phi^{\mathrm{o,i}} - \phi^{\mathrm{i,o}})\right] \\
&\quad + r^{\mathrm{s}}\left(-\frac{1}{T^{\mathrm{s}}}\Delta_n G^{\mathrm{s}}\right) \\
&= J_q^{\mathrm{i,o}}\Delta_{\mathrm{i,s}}\frac{1}{T} + J_q^{\mathrm{o,i}}\Delta_{\mathrm{s,o}}\frac{1}{T} + \sum_{j=1}^{n} J_j^{\mathrm{i,o}}\left(-\Delta_{\mathrm{i,s}}\frac{\mu_j}{T}\right) \\
&\quad + \sum_{j=1}^{n} J_j^{\mathrm{o,i}}\left(-\Delta_{\mathrm{s,o}}\frac{\mu_j}{T}\right) + j\left(-\frac{1}{T^{\mathrm{s}}}\Delta_{\mathrm{i,o}}\phi\right) \\
&\quad + r^{\mathrm{s}}\left(-\frac{1}{T^{\mathrm{s}}}\Delta_n G^{\mathrm{s}}\right)
\end{aligned} \qquad (9.10)
$$

In the last identity we introduced the notation for transport into and across the surface that was illustrated in Fig. 9.3. Some components are only present on one side of the surface. The flux of these components on the other side is then zero. This reduces the number of contributions to the excess entropy production.

We can also use the measurable heat flux as a variable:

$$J_q(x,t) = J_q'(x,t) + \sum_{j=1}^{n} H_j(x,t)J_j(x,t) \qquad (9.11)$$

Under stationary state conditions at an electrode, the reaction rate is proportional to the electric current density, $r^{\mathrm{s}} = j/zF$, where $z$ is

the number of electrons involved in the reaction. This and Eq. (9.11)
gives

$$
\sigma^s = J_q^{\prime i,o} \Delta_{i,s} \frac{1}{T} + J_q^{\prime o,i} \Delta_{s,o} \frac{1}{T} + \sum_{j=1}^{n} J_j^{i,o} \left[ -\frac{1}{T^s} \Delta_{i,s} \mu_{j,T}(T^s) \right]
$$

$$
+ \sum_{j=1}^{n} J_j^{o,i} \left[ -\frac{1}{T^s} \Delta_{s,o} \mu_{j,T}(T^s) \right] + j \left[ -\frac{1}{T^s} \left( \Delta_{i,o} \phi + \frac{\Delta_n G^s}{zF} \right) \right]
$$

$$(9.12)$$

The meaning of the term $-j\Delta_{i,o}\phi$ comes from the energy balance as
mentioned before, Eq. (9.7), and the flux equations that follow from
Eq. (9.12). The product is the electric power delivered to the interface when the current $j$ is passing the cell. It leads to changes in
the other terms. In the absence of an external circuit and an electric
current, the term disappears from the entropy production.

The excess entropy production is independent of the frame of reference, and so are the notions, reversible and irreversible. They are
in other words invariant under a coordinate transformation. We may
therefore convert all fluxes and conjugate forces from any frame of
reference, to the surface frame of reference and back, without chang-
ing the entropy production in the different phases, $\sigma^i$, $\sigma^s$ and $\sigma^o$.

## Exercise 9.3.1

*Consider stationary state evaporation of A from a solution. The flux
is $1.6 \ 10^{-6} \ mol.m^{-2}.s^{-1}$. The other component does not evaporate.
The concentrations close to the surface are $c_A^i = 0.05 \ kmol.m^{-3}$ on
the liquid side, and $c_A^o = 0.01 \ kmol.m^{-3}$ on the gas side. What is the
excess entropy production in the surface?*

- **Solution:** At stationary state, there is no accumulation of A in
  the surface and $J_A^{i,o} = J_A^{o,i}$. When the temperature is constant,
  the excess entropy production is

$$
\sigma^s = J_A^{i,o} \left[ -\frac{1}{T^s}(\mu_A^s - \mu_A^{i,o}) \right] + J_A^{o,i} \left[ -\frac{1}{T^s}(\mu_A^{o,i} - \mu_A^s) \right]
$$

$$
= J_A \left[ -\frac{1}{T^s} \left( \mu_A^{o,i} - \mu_A^{i,o} \right) \right] = J_A R \ln \frac{c_A^i}{c_A^o} = 2.1 \times 10^{-5} \ W.K^{-1}.m^{-2}
$$

This excess entropy production is not a direct function of the temperature. In order to compare the surface's excess entropy production to the one in homogeneous phases, we divide this value with the thickness of the surface, which is order of magnitude $10^{-9}$ m. The result is very large, 21 kW.K$^{-1}$.m$^{-3}$, compared for instance to the values calculated in Exercise 3.3.8. The interface is, relatively speaking, often a large source of entropy production.

## Exercise 9.3.2

*Derive Eq. (9.12) from Eq. (9.10).*

- **Solution:** We have

$$
\Delta_{i,s}\left(\frac{\mu_j}{T}\right) = \frac{\mu_j^s(T^s)}{T^s} - \frac{\mu_j^{i,o}(T^{i,o})}{T^{i,o}}
$$

$$
= \frac{\mu_j^s(T^s)}{T^s} - \frac{\mu_j^{i,o}(T^s)}{T^s} + \frac{\mu_j^{i,o}(T^s)}{T^s} - \frac{\mu_j^{i,o}(T^{i,o})}{T^{i,o}}
$$

$$
= \frac{1}{T^s}\Delta_{i,s}\mu_{j,T} + \left(\frac{\partial}{\partial T}\frac{\mu_j^{i,o}}{T}\right)_{T=T^{i,o}} (T^s - T^{i,o})
$$

$$
= \frac{1}{T^s}\Delta_{i,s}\mu_{j,T} - \left(\frac{1}{T^{i,o}}S_j^{i,o} + \frac{\mu_j^{i,o}}{(T^{i,o})^2}\right)(T^s - T^{i,o})
$$

$$
= \frac{1}{T^s}\Delta_{i,s}\mu_{j,T} + \left(S_j^i + \frac{\mu_j^{i,o}}{T^{i;o}}\right)T^s\Delta_{i,s}\left(\frac{1}{T}\right)
$$

$$
= \frac{1}{T^s}\Delta_{i,s}\mu_{j,T} + H_j^{i,o}\left(\frac{T^s}{T^{i,o}}\right)\Delta_{i,s}\left(\frac{1}{T}\right)
$$

We can assume that, $T^s \approx T^i$, so that

$$
\Delta_{i,s}\left(\frac{\mu_j}{T}\right) = \frac{1}{T^s}\Delta_{i,s}\mu_{j,T} + H_j^{i,o}\Delta_{i,s}\left(\frac{1}{T}\right)
$$

By substituting this result, and the analogous result for the other side of the surface, for all components into Eq. (9.10), and using the definition of the measurable heat fluxes on both sides, we obtain Eq. (9.12).

Equation (9.12) uses the notation for change in the chemical potential from a value in the homogeneous phases close to the surface, in the point a, to a value in the homogeneous phase on the other side, at point b, *cf.* Fig. 9.3. At the temperature of the surface, we have

$$\Delta_{i,s}\mu_{j,T}(T^s) = \mu_j^s(T^s) - \mu_j^{i,o}(T^s)$$

$$\Delta_{s,o}\mu_{j,T}(T^s) = \mu_j^{o,i}(T^s) - \mu_j^s(T^s) \qquad (9.13)$$

Subscript $T$ means that the chemical potential difference is evaluated at the surface temperature. Each difference is written as the value to the right minus the value to the left. This choice gives the differences the same sign as the gradients in the homogeneous phases, for increasing or decreasing variables. The subscripts of $\Delta$ refer to the two locations between which the difference is taken.

It is interesting to compare Eqs. (9.10) and (9.12) with the equations we have for the homogeneous phase. The gradients of the temperature and the chemical potentials have been replaced by differences of these variables into and out of the surface, the fluxes are the same. The gradient of the electric potential has been replaced by a difference across the surface. These replacements represent the discrete nature of the surface and are typical for transport between different homogeneous parts of a system [107].

Most of the contributions to the excess entropy production of a planar, isotropic surface, as given above, are due to fluxes of heat, mass and charge in the homogeneous phases outside the surface. The contribution from the chemical reaction can, however, be a specific *surface* contribution, like in heterogeneous catalysis. At the surface, the system is no longer isotropic in the direction normal to the surface, so that all the normal flux components, unlike in the homogeneous phase, are scalar under rotations and reflections in the plane of the surface. We can therefore have coupling between a chemical or electrochemical reaction and the normal of the fluxes of heat, mass and charge at a surface. Mass transport fueled by energy from a chemical reaction is called active transport in biology [16, 82]. *The coupling that takes place between scalar fluxes at the surface, make transports in heterogeneous systems different from transports in a homogeneous system.*

We need a *common* frame of reference for the fluxes, when we want to describe transport in heterogeneous materials. We have discussed above that, the electric current density, $j$, is constant in electroneutral systems and independent of the frame of reference. Mass fluxes, on the other hand, depend on the frame of reference. The frame of reference that gives a simple realistic description of the surface, is the surface itself. In this frame of reference the observer moves along with the surface. The natural choice of the frame of reference for a heterogeneous system is therefore the *surface frame of reference*. All expressions for the entropy production of the surface, contain fluxes relative to this frame of reference. In heterogeneous systems, we will use this frame of reference *also for the adjacent homogeneous phases.* For a definition of the various frames of reference, and transformations between them, we refer to Section 3.4 [12, 31].

### Exercise 9.3.3

*Consider the special case that one component $k$ is transported through an isothermal surface. The densities of the other components, the internal energy, the molar surface area and the polarization densities are all independent of the time. Show that the excess entropy production is given by*

$$\sigma^s = J_k^{i,o} \left[ -\frac{1}{T^s}(\mu_k^s - \mu_k^{i,o}) \right] + J_k^{o,i} \left[ -\frac{1}{T^s}(\mu_k^{o,i} - \mu_k^s) \right]$$

- **Solution:** In this case, Eq. (9.9) reduces to

$$T^s \frac{ds^s}{dt} + \mu_k^s \frac{d\Gamma_k}{dt} = 0$$

The rate of change of the entropy is therefore given by

$$\frac{ds^s}{dt} = -\frac{\mu_k^s}{T^s} \frac{d\Gamma_k}{dt} = -\frac{\mu_k^s}{T^s} \left( J_k^{i,o} - J_k^{o,i} + r_k^s \right)$$

where we substituted Eq. (9.4). Setting $r_k^s = 0$ one obtains

$$\frac{ds^s}{dt} = -\frac{\mu_k^s}{T^s}(J_k^{i,o} - J_k^{o,i})$$

$$= -\frac{\mu_k^{i,o}}{T^s} J_k^{i,o} + \frac{\mu_k^{o,i}}{T^s} J_k^{o,i} - J_k^{i,o} \left( \frac{\mu_k^s}{T^s} - \frac{\mu_k^{i,o}}{T^s} \right) - J_k^{o,i} \left( \frac{\mu_k^{o,i}}{T^s} - \frac{\mu_k^s}{T^s} \right)$$

$$= J_s^{i,o} - J_s^{o,i} + J_k^{i,o} \left[ -\frac{1}{T^s}(\mu_k^s - \mu_k^{i,o}) \right] + J_k^{o,i} \left[ -\frac{1}{T^s}(\mu_k^{o,i} - \mu_k^s) \right]$$

By comparing this equation with Eq. (9.2), we can identify the wanted excess entropy production.

**Exercise 9.3.4**

*Consider the special case that only heat is transported. The excess entropy production is given by*

$$\sigma^s = J_q'^{i,o}\left(\frac{1}{T^s} - \frac{1}{T^{i,o}}\right) + J_q'^{o,i}\left(\frac{1}{T^{o,i}} - \frac{1}{T^s}\right)$$

*What are the flux-force relations?*

- **Solution:** The flux-force relation that follow from the entropy production are

$$J_q'^{i,o} = l_{ii}\left(\frac{1}{T^s} - \frac{1}{T^{i,o}}\right) + l_{io}\left(\frac{1}{T^{o,i}} - \frac{1}{T^s}\right)$$

$$J_q'^{o,i} = l_{oi}\left(\frac{1}{T^s} - \frac{1}{T^{i,o}}\right) + l_{oo}\left(\frac{1}{T^{o,i}} - \frac{1}{T^s}\right)$$

These equations express a possibility for coupling between the incoming and outgoing heat fluxes.

## 9.4   Stationary state evaporation and condensation

Evaporation or condensation takes place constantly in our environment due to temperature and pressure changes. The most important is the life cycle of water. Also the industry needs to model evaporation and condensation, in distillation towers and other separators. Expressions for the mass flux across phase boundaries are in both cases at the center of interest. Non-equilibrium thermodynamics teaches us that a mass flux does not take place without a heat flux; and indeed we know very well that the addition of heat leads to evaporation. The theory provides a relation between the two fluxes through the Onsager relations.

For stationary evaporation and condensation, the expression for the excess entropy production, Eq. (9.10), simplifies greatly. Only two flux-force pairs remain after the introduction of stationary state conditions, $J_q^{i,o} = J_q^{o,i} \equiv J_q$ and $J^{i,o} = J^{o,i} \equiv J$ [101]:

$$\sigma^s = J_q \Delta_{i,o} \frac{1}{T} + J \left( -\Delta_{i,o} \frac{\mu}{T} \right) \tag{9.14}$$

The resulting linear force-flux relations are

$$\Delta_{i,o} \frac{1}{T} = r_{qq}^s J_q + r_{q\mu}^s J$$

$$-\Delta_{i,o} \frac{\mu}{T} = r_{\mu q}^s J_q + r_{\mu\mu}^s J \tag{9.15}$$

This set of equations is convenient when the total heat flux is constant. Only three coefficients need be determined, since the Onsager symmetry relations give $r_{q\mu}^s = r_{\mu q}^s$. These coefficients can be derived from molecular simulations.

When the measurable heat flux is the wanted variable, we must choose the side we refer this variable to, as this heat flux is not continuous at the surface. Consider the measurable heat flux on the o-side as variable and eliminate the other using the energy balance. The excess entropy production in Eq. (9.12) becomes:

$$\sigma^s = J_q^{'o,i} \Delta_{i,o} \frac{1}{T} + J \left( -\frac{1}{T^i} \Delta_{i,o} \mu(T^i) \right) \tag{9.16}$$

and the resulting linear force-flux relations are

$$\Delta_{i,o} \frac{1}{T} = r_{qq}^s J_q^{'o,i} + r_{q\mu}^{s,o} J$$

$$-\frac{1}{T^i} \Delta_{i,o} \mu(T^i) = r_{\mu q}^{s,o} J_q^{'o,i} + r_{q\mu}^{s,o} J \tag{9.17}$$

The resistivity coefficients of this matrix can be derived from experiments. Again, only three coefficients are independent. The sets of equations apply to any stationary-state phase transition of one component. The sets apply for flat as well as curved interfaces. The resistivities are, in general, functions of the surface intensive variables; i.e. the surface temperature, curvature and excess density.

The coefficients of Eq. (9.17) have been determined from kinetic theory [31, 101], which applies to hard spheres. Data for real systems

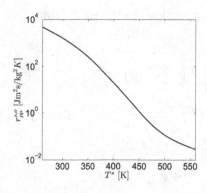 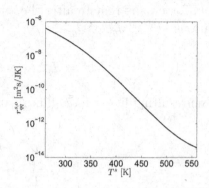

(a) Main resistivity to mass transfer.

(b) Main resistivity to heat transfer.

Figure 9.6: Resistivities for evaporation or condensation of water as a function of temperature. Reprinted with permission from *Phys. Rev. E* [110].

are lacking, but first progress has been made [31, 108, 109, 110, 101]. Wilhelmsen *et al.* [109, 110] found the resistivities for argon-like particles (particles that interact with a Lennard-Jones potential) and for water, using a combination of experimental results, molecular dynamics simulations, and square gradient theory. Klink *et al.* [108] combined classical density functional theory with the perturbed-chain statistical associating fluid theory (PC-SAFT). The equation of state was used to first determine the density and enthalpy profiles through the interface. Coefficients were derived, not only for pure, but also for two-component mixtures. These studies can provide routes to interface transfer coefficients in real systems. The results for flat surfaces of water are shown as a function of temperature in Figs. 9.6 and 9.7.

The two main resistivity coefficients and the coupling coefficient vary by orders of magnitude as a function of $T$. All coefficients decrease with temperature, as expected (they vanish at the critical point). The heat of transfer, equal to minus the ratio of the coupling coefficient to the main coefficient for heat transfer, is almost negligible near 300 K, but becomes non-negligible as the temperature increases, see Fig. 9.7. It was established that the coefficients of Eq. (9.17) also depend strongly on the surface curvature when the radius of the bubble or droplet is in the nanometer range [110].

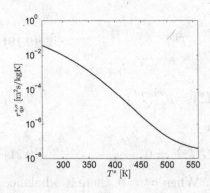 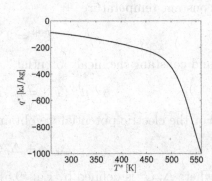

(a) Resistivity for coupled trans-
    fer of heat and water.

(b) Heat of transfer for water
    evaporation.

Figure 9.7: Coupling coefficient for coupled transfer of heat and mass (left) and the heat of transfer (right). Reprinted with permission from *Phys. Rev. E* [110].

There is a relation between the sets of coefficients, Eqs. (9.17) and (9.15), due to the invariance of the entropy production. Using this and the Gibbs-Helmholtz equation, we obtain

$$r_{q\mu}^{s,o} = r_{\mu q}^{s,o} = r_{q\mu}^{s} + H^{o}\, r_{qq}^{s}$$
$$r_{\mu\mu}^{s,o} = r_{\mu\mu}^{s} + 2H^{o}\, r_{q\mu}^{s} + (H^{o})^{2}\, r_{qq}^{s} \tag{9.18}$$

The sets of coefficients are related by the enthalpy of the o-phase. Choosing this enthalpy, we can compute the set that belongs to the total energy flux, from the results in Figs. 9.6 and 9.7. The results can be compared to those obtained from molecular dynamics simulation data. Sets of coefficients were also determined for the freeze crystallization of ice [111, 112].

## 9.5 Equilibrium at the electrode surface. Nernst equation

The expression for the excess entropy production contains useful information about reversible electrode processes. For these processes, the excess entropy production is zero. By setting $\sigma^{s} = 0$ in Eq. (9.12) we find the equilibrium conditions of the electrode surface. These are constant temperature

$$T^{o,i} = T^s = T^{i,o}, \tag{9.19}$$

and constant chemical potential

$$\mu_j^{i,o}(T^s) = \mu_j^s(T^s) = \mu_j^{o,i}(T^s) \tag{9.20}$$

For the electric potential we obtain

$$\Delta_{i,o}\phi + \Delta_n G^s / zF = 0 \tag{9.21}$$

where $\Delta_n G^s$ is defined by Eq. (9.5). When $\sigma^s = 0$, there is a balance of forces. The electrode potential jump is

$$\Delta_{i,o}\phi = -\Delta_n G^s / zF \tag{9.22}$$

This is the Nernst equation. The equation describes the reversible measurement of the electromotive force.

The reaction Gibbs energy for a single electrode reaction depends on a reference state, however. Take as an example the hydrogen electrode. The electrode reaction is

$$\frac{1}{2}H_2(g) \rightleftharpoons H^+ + e^- \tag{9.23}$$

In general, when $j \approx 0$,

$$\Delta_{a,e}\phi(H_2) = \frac{1}{2F}\mu_{H_2} \tag{9.24}$$

The reaction is used as a zero reference in tables for standard reduction potentials. At standard conditions the hydrogen pressure is 1 bar. The potential drop due to the reaction, is set as zero at 1 bar and 298 K. From Eq. (9.24) we have:

$$\Delta_{a,e}\phi^0(H_2) = \frac{1}{2F}\mu_{H_2}^0 \equiv 0 \tag{9.25}$$

At any other pressure, the potential drop across the surface becomes

$$\Delta_{a,e}\phi(H_2) = \frac{1}{2F}\mu_{H_2} = \frac{1}{2F}(\mu_{H_2}^0 + RT \ln p_{H_2}/p^0) \tag{9.26}$$

The value of $\Delta_{a,e}\phi(H_2)$ for the anode in the polymer electrolyte fuel cell, where the electrolyte is an acid membrane, is $-0.23$ V at 340 K and a hydrogen pressure of 0.76 bar.

**Remark 10** *The value of $\Delta_{i,o}\phi$ depends on the standard state that is chosen for the chemical potential of the neutral components. It is thus not an absolute quantity. When the potential difference is calculated between two electrodes, this arbitrariness must disappear, as the measured potential difference is absolute. Care must be taken to use the same standard states for both electrodes, see Appendix A.3.*

The description using absorptions of neutral components is equivalent to a description using charged and uncharged components. Equivalence means $\Delta_{i,o}\phi + \Delta_n G^s/F = \Delta_{i,o}\psi + \Delta_r G^s/F$ where $\Delta_r G^s$ is the reaction Gibbs energy, *cf.* Eq. (9.4). The Nernst equation obtains its more usual form with the Maxwell potential difference [31].

$$\Delta_{i,o}\psi = -\frac{1}{F}\Delta_r G^s = -\frac{1}{F}\left(\mu_{H^+}^0 + \mu_{e^-}^0 - \frac{1}{2}\mu_{H_2}^0\right) \qquad (9.27)$$

Standard reduction potentials are constructed on this basis, with the electrolyte standard state of a 1 molar solution of protons. The difference $\Delta_{i,o}\phi$ is the Maxwell potential difference $\Delta_{i,o}\psi$ plus the sum of the chemical potentials of the charge carriers divided by Faraday's constant.

**Remark 11** *The advantage of using $\Delta_{i,o}\phi$ is its relation to measurable quantities, as discussed by Guggenheim [113, 114]. The Maxwell potential difference includes chemical potentials of ions, which cannot be measured.*

## 9.6 Stationary states at electrode surfaces. The overpotential

Out of equilibrium, when the electric current density is large, temperature and concentration differences may arise near the surface, and other terms in the entropy production may become important [100]. All forces and fluxes in Eq. (9.12) are coupled, and the situation becomes complex. In the stationary state, some simplifications can be obtained. It follows from Eqs. (9.4) and (9.7) that

$$J_j^{i,o} - J_j^{o,i} + \nu_j r^s = 0$$
$$J_q^{i,o} - J_q^{o,i} - j\Delta_{i,o}\phi = 0 \qquad (9.28)$$

when all components are present on both sides of the surface. The mass and total heat fluxes on both sides of the surface are no longer independent in the stationary state. We choose to eliminate, for instance, the fluxes on the o-side in Eq. (9.10). This gives

$$\sigma^s = J_q^{i,o}\Delta_{i,o}\frac{1}{T} + \sum_{j=1}^{n} J_j^{i,o}\left(-\Delta_{i,o}\frac{\mu_j}{T}\right) + j\left[-\frac{1}{T^o}(\Delta_{i,o}\phi + \frac{\Delta_n G^o}{zF}\right]$$

(9.29)

Alternatively, we can eliminate the fluxes on the i-side and obtain a similar equation. When a component is only present on one side of the surface, we keep the expression in Eq. (9.29) for that component.

Away from equilibrium, the other forces in Eq. (9.29) contribute to $\Delta_{i,o}\phi$. We define the overpotential in accordance with Newman [115]

$$\eta \equiv \Delta_{i,o}\phi + \Delta_n G^s/zF$$

(9.30)

to obtain the effective electrochemical driving force. The overpotential is defined as a positive quantity, which means that we must take this into account. Flux equations for $J_q^{\prime i,o}$, $J_q^{\prime o,i}$, $J_j^{i,o}$, $J_j^{o,i}$ follow as usual from the entropy production. We do not give further details on this here.

When the electric current density is small, the force-flux relations are linear, and

$$\eta = l_T^{s,i}\Delta_{i,s}T + l_T^{s,o}\Delta_{s,o}T + \sum_{j=1}^{n} l_\mu^{s,i}\Delta_{i,s}\mu_{j,T}(T^s) + \sum_{j=1}^{n} l_\mu^{s,o}\Delta_{s,o}\mu_{j,T}(T^s) - r_\phi j$$

(9.31)

In addition to the concentration overpotential represented by, $l_\mu^{s,i}\Delta_{i,s}\mu_{j,T}(T^s)$ and $l_\mu^{s,o}\Delta_{s,o}\mu_{j,T}(T^s)$, there can be contributions to the overpotential from thermal driving forces, $l_T^{s,i}\Delta_{i,s}T$ and $l_T^{s,o}\Delta_{s,o}T$.

Experiments show that the reaction overpotential is linear in the current density only when very close to equilibrium. To find an expression for $\eta$ that applies to the non-linear regime, we use the method described in Chapter 7 [81], see also [116]. By introducing a probability distribution for the reaction along an internal coordinate, we derive the Butler-Volmer equation:

$$j = j_0 \left[ \exp^{(1-\alpha)\eta F/RT} - \exp^{-\alpha\eta F/RT} \right] \qquad (9.32)$$

Here $j_0$ is the exchange current density at equilibrium, and the transfer factor $\alpha$ gives the position of the activation energy barrier. The Butler-Volmer equation has the same basis as the Nernst equation. The equation applies for isothermal conditions.

## 9.7  Concluding remarks

We have seen in this chapter how the entropy production for a surface can be constructed. Doing this, we obtain dynamic boundary conditions for transport of heat, mass and charge into and across the surface. This is important as the surface is often the source of large entropy production. This is the case during phase transitions and during electrode processes.

The surface need not be only of molecular dimensions. The next chapter provides an example of transport through a thick surface; a membrane.

# Chapter 10

# Transport through membranes

*We describe coupled transport of heat, mass and charge in membranes. The coupled transport of water and heat in hydrophobic membranes can be used to produce pure water and/or power from waste heat. The coupled transport of charge and mass transport in ion-exchange membranes can be used to produce power by mixing salt and brackish water.*

## 10.1 Introduction

Membranes play important roles as separators. A membrane, unlike a solution, can sustain a pressure difference, meaning that the pressure can play a role as a driving force. Water can undergo a phase change entering the membrane. A phase change is often connected to a heat effect. This means that mass transport is coupled to heat transport. In ion-exchange membranes, mass transport means that there is also charge transport. Electric work can be produced or added to systems of membranes and electrolyte solutions.

There are large amounts of available heat around us, in the process industry, from geothermal sources, or from the sun. These can be explored as energy sources for separation. For instance, it is interesting to consider the thermal driving force for the production of pure water. Pure water is increasingly being produced from salt or brackish water. A membrane distillation process for the production of clean water was already proposed in 1967 [117]. It has recently been proposed that pure water production can occur together with power production [118, 119].

The purpose of this chapter is to describe transport in membranes. In the cases mentioned, the coupling coefficients are large. Systems with membranes are heterogeneous, *cf.* Chapter 9. They consist of bulk and surface parts. We can regard them as a combination of three- and two-dimensional subsystems, with entropy production in all subsystems. We limit ourselves here to describe examples where the whole membrane can be treated as one interface. The description becomes discrete (discontinuous), *cf.* Eq. (10.1). From the entropy production of the membrane, we shall find flux equations for the heterogeneous system. For treatment of more advanced cases, see [31].

We shall examine cases where thermal, chemical and electrical driving forces appear. In the previous chapter, we gave the excess entropy production of a surface when the measurable heat flux on the o-side was chosen. The chemical driving force must then be evaluated at the temperature of the i-side. The entropy production of the membrane is:

$$\sigma^m = J_q'^o \Delta \left(\frac{1}{T}\right) + \sum_j J_j \left(-\frac{1}{T^i}\Delta\mu_j(T^i)\right) + j\left(-\frac{1}{T^i}\Delta\phi\right) \quad (10.1)$$

For terminology, see Chapter 9. The summation is carried out over the independent neutral components. We shall deal with two components or less; water and possibly an electrolyte. Both components can move relative to the membrane; the natural frame of reference for the fluxes. In this choice of components, we are in line with the practical choice made by Katchalsky and Curran, and Førland and coworkers [15, 19]. What is special is that the theory, as extended to surfaces [31], is used to obtain a precise definition of the conjugate fluxes and forces in Eq. (10.1). This is relevant for thermal osmosis and electro-osmosis, see below.

## 10.2 Osmosis

In *osmosis*, water is moving down its chemical potential gradient. In the simplest case, only water is moving, driven by the difference in chemical potential across the membrane. The water flux, $J_w$, is proportional to $\Delta\mu_w$.

$$J_w = -L\Delta\mu_w \qquad (10.2)$$

where $L$ is the water permeability. The chemical potential has a concentration- and a pressure dependent term; $\mu_w = \mu_w^c + V_w p$, where $V_w$ is the molar volume of water and $p$ is the pressure, see Appendix A.2. This gives $\Delta\mu_w = \Delta\mu_w^c + V_w\Delta p$. A concentration-dependent term arises when there is a concentration difference of solute.

We observe a net water flux, *osmosis*, until there is equilibrium for water across the membrane. At equilibrium, $\Delta\mu_w = 0$ and $\Delta p = -\Delta\mu_w^c/V_w$. In order to introduce the properties of the solute, we use Gibbs-Duhem's equation for the solution; $c_w d\mu_w^c = -c_j d\mu_j^c$. The expression for $\Delta p$ is obtained from $dp = (c_j/c_w V_w)d\ \mu_j^c$. We integrate over water solutions in equilibrium with the membrane at any location in the membrane (Scatchard's assumption) [120]. When the solutions are ideal, $c_w V_w \approx 1$, we obtain Van't Hoff's equation

$$\Pi \equiv \Delta p_{J_w=0} = c_j\Delta\mu_j^c \approx RT\Delta c_j \qquad (10.3)$$

For a concentration difference of 0.3 kmol.m$^{-3}$ (the solute concentration in blood) at a temperature $T = 300$ K, we compute $\Pi = 7.5$ bar. Water accumulates on the side where the solute has the highest concentration, increasing the chemical potential of water on this side.

**Exercise 10.2.1** *Water can be purified by reverse osmosis, by applying high pressure to the side with low solute concentration. Compute the minimum pressure needed at 300 K when the concentration of the salt solution is 0.3 kmol.m$^{-3}$.*

- **Solution:** The salt solution is on the left-hand side of the membrane, the pure water is on the right-hand side of the membrane. The positive direction of transport is taken from left to right. Van't Hoff's equation contains the number of solute particles, here 0.6 kmol.m$^{-3}$, if we assume that the salt is fully dissociated. The $\Delta c_k$ is negative. The minimum pressure is obtain by introducing this value for $\Delta c_k$ in Eq. (10.3).

$$\Pi = -8.31 \text{ J.K}^{-1}.\text{mol}^{-1}300 \text{ K}(0.6 \text{ kmol.m}^{-3}) = -15 \text{ bar}$$

We need to apply more than 15 bar to the concentrated salt solution, in order to force water through the membrane against its chemical potential.

## 10.3   Thermal osmosis

*Thermal osmosis* means that a water flux arises due to a thermal driving force. Consider a membrane surrounded by two solutions. There are two driving forces, one for heat and one for water transport. The difference in the water chemical potential can have a contribution from the presence of salt on the feed side. For the time being we neglect salt transport. The chemical force is the main driving force for water. But the thermal driving force will also have an effect on mass transport through the coupling coefficient. The flux equations are:

$$J_q'^\text{o} = L_{qq}\Delta_{\text{i,o}}\left(\frac{1}{T}\right) - L_{qw}\frac{1}{T^\text{i}}\Delta\mu_w(T^\text{i})$$

$$J_w = L_{wq}\Delta_{\text{i,o}}\left(\frac{1}{T}\right) - L_{ww}\frac{1}{T^\text{i}}\Delta\mu_w(T^\text{i}) \qquad (10.4)$$

We relate $L_{qq}/(T^\text{i}T^\text{o})$ to the Fourier conductivity, while $L_{ww}/T^\text{i}$ is the water permeability, $L$. The coefficient ratio is the heat of transfer

$$q^* = \frac{L_{qw}}{L_{ww}} \qquad (10.5)$$

The coupling coefficient $L_{qw} = L_{wq}$ can be found at Soret equilibrium, when $J_w = 0$. When $J_w = 0$, the thermal force balances the chemical force. The chemical driving force is equal to $-V_w\Delta p$ in the absence of salt. The heat of transfer of the membrane can be large, similar to values observed for phase transitions, *cf.* Chapter 9. The sign of the heat of transfer can vary and depends on the membrane's properties.

**Exercise 10.3.1** *Water can accumulate a pressure reservoir on one side of a nanoporous membrane, if the membrane is exposed to a temperature difference. Calculate the maximum pressure rise, for a temperature difference of $\Delta T = 6.5$ K at an average temperature of $T = 300$ K. The molar volume of water is $V_w = 18 \times 10^{-6}$ m³.mol⁻¹ and $q^* = -2$ kJ.mol⁻¹.*

- **Solution:** From Eq. (10.4) and the given constants, we find from Eq. (10.3) that

$$(\Delta p)_{J_w=0} = -\frac{1}{V_w}\frac{L_{qw}}{L_{ww}}\frac{\Delta T}{T} = -\frac{q^*}{V_w}\frac{\Delta T}{T}$$

The average temperature is used for $T$. A negative sign of the heat of transfer means that pressure builds on the high-temperature side, $\Delta p_{J_w=0} > 0$. From this we compute $\Delta p_{J_w=0} = 24$ bar. Water accumulates on the high-temperature side. The pressure reservoir can be used to run a turbine, if it can be maintained fast enough.

### 10.3.1 Water and power production

Consider the principle of a recent invention [118, 119] for the simultaneous production of pure water and power; the MemPower process. A sketch of the system is shown in Fig. 10.1. Water enters the membrane as liquid on one side, then evaporates and diffuses to the other side where the temperature is so low that it condenses. This is observed in hydrophobic membranes. Water is transported as vapor through the membrane pores driven by a thermal force. The water

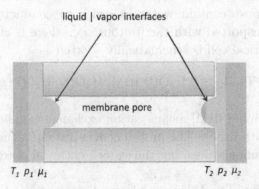

Figure 10.1: A schematic illustration of the hydrophobic membrane pore, where water is transported across a temperature gradient in the vapor phase.

flux leads to transport of heat, but also to a pressure rise on the receiving side. The pressure reservoir can in turn be used for power production by a turbine as explained in exercise 10.3.1. In this way, one can produce not only pure water, but also power from waste heat.

The equations in Eq. (10.4) can be used to describe the overall membrane processes. This was done by Keulen *et al.* [121]. They first considered the liquid – vapor interface on the left-hand side, and then the vapor – liquid interface on the right-hand side. The coefficients for evaporation of water presented in Section 9.4, gave reasonable agreement with experimental observations of the mass flux.

## 10.4   Electro-osmosis at constant temperature

When we apply an electric driving force to a membrane, we can observe *electro-osmosis*; transport of water by an electric current.

Consider as an example, a cation exchange membrane with anionic sites, $M^-$, like the Nafion membrane of the common fuel cell. The membrane conducts protons. In our case, a hydrogen electrode on the left-hand side supplies protons, while the same electrode on the right-hand side removes protons. The reactions take place on Pt grains embedded in carbon, in contact with the membrane. The membrane must contain water in order to conduct protons well. Water is transported with the protons; *i.e.* there is electro-osmosis. The symmetrical cell is schematically written as

$$(p_1, T) \ H_2 \ (g), \ H_2O(l) \ |HM| \ H_2O(l), \ H_2(g) \ (p_2, T)$$

where the symbol |HM| means cation exchange membrane in proton form filled with water. The proton concentration in the membrane is fixed by the anionic sites, but the water's chemical potential can vary.

The cell potential under reversible conditions has contributions from the electrodes and the membrane. We deal with each contribution separately. In the overall description they are added.

### 10.4.1 Contributions from the electrodes

The entropy production of an electrode surface was given in Chapter 9. From Section 9.4 we obtain the contribution to the cell potential by applying Eq. (9.26) to both electrodes, and adding the results

$$\Delta_{el}\phi = -\frac{RT}{2F}\ln\frac{p_2}{p_1} \tag{10.6}$$

### 10.4.2 Contributions from the membrane

The entropy production of the membrane has two terms, one due to movement of water and one due to movement of charge. The flux-force equations for the membrane become

$$J_w = -L_{\mu\mu}\Delta\mu_w - L_{\mu\phi}\Delta\phi$$
$$j = -L_{\phi\mu}\Delta\mu_w - L_{\phi\phi}\Delta\phi \tag{10.7}$$

The flux of protons multiplied by Faraday's constant, $F$, is equal to the electric current, $j$, and is not an independent flux. Since the cell is isothermal, we have dropped subscript $T$ in $\mu_w$, and included the factor $1/T^i$ into the conductivities.

The electric resistivity of the membrane is

$$r \equiv \frac{1}{L_{\phi\phi}} \tag{10.8}$$

The coefficient $L_{\mu\mu}$ describes transport of water when the cell is short-circuited ($\Delta\phi = 0$). The transference coefficient of water is defined as:

$$t_w = F\left(\frac{J_w}{j}\right)_{\Delta\mu_w=0} = F\frac{L_{\mu\phi}}{L_{\phi\phi}} \tag{10.9}$$

The coefficient can be found by measuring the water flux, $J_w$, as a function of the electric current density in the membrane.

When the water concentration is the same on the two sides, but a hydrostatic pressure difference exists, $\Delta\mu_w = V_w\Delta p$. We solve Eq. (10.7) for $j = 0$, using $L_{\phi\mu} = L_{\mu\phi}$ and obtain:

$$\Delta\phi = -\frac{t_w}{F}V_w\Delta p \tag{10.10}$$

We see from Eq. (10.10) that we can measure the electric potential difference due to a pressure difference, and find $t_w$ [122]. With liquid water on the sides of the membrane, $t_w = 2.6$ [122], and $V_w = 18 \times 10^{-6}$ m$^3$.mol$^{-1}$, a pressure difference of 1 bar generates an electromotive force of $-9$ mV at 300 K. A very large pressure difference is needed to generate a potential difference near that of a battery.

There is a relation between the two experiments, given by the Onsager relation

$$\left(\frac{J_w}{j}\right)_{\Delta\mu_w=0} = -\left(\frac{\Delta\phi}{\Delta p}\right)_{j=0} \tag{10.11}$$

Equation (10.11) is known as Saxén's relation. Volume or water flow by electric current is called *electro-osmosis* (see above), while an electric potential generated by pressure difference is called a *streaming potential*.

Following the same procedure as in Chapter 4's Eq. (4.52), we find $D_w$, the diffusion coefficient of water in the membrane from $J_w = -D_w\Delta c_w$ when $j = 0$.

In the presence of an electric current, the water flux is either

$$J_w = -D_w\Delta c_w + \frac{t_w}{F}j \tag{10.12}$$

or

$$J_w = -L_p\Delta p + \frac{t_w}{F}j \tag{10.13}$$

where we replaced the first term on the right-hand side by the Darcy law. The coefficient $L_p = L_{\mu\mu}V_w$ is the hydraulic permeability of the membrane. Pure diffusion of water is described by setting $j = 0$. Osmosis (the first terms on the right-hand side) is superimposed on the electro-osmotic transport (the last term on the right-hand side).

Electro-osmosis also takes place in nature. It has, for instance, been documented [123] that the water content inside an animal's eye, is regulated by a flux of sodium ions that drags water back into the eye through a leaky epithelial cell layer that covers the outside of the cornea (the external eye). The current density used to move water can be as high as 600 A.m$^{-2}$ [123]. Electro-osmosis has also been shown to speed-up the drying of concrete.

**Exercise 10.4.1** *Electro-osmosis has been called electrochemical pumping. The transport of water due to $t_w j/F$ leads, with $t_w > 0$, to transport along with positive charge carriers. In a polymer electrolyte fuel cell membrane, where protons carry charge, we need to replenish water at the site where protons enter the membrane due to this effect. A concentration gradient of water will build across the membrane. Compute the maximum gradient in water concentration, when $t_w = 2.6$, $D_w = 2 \times 10^{-10}$ $m^2.s^{-1}$ and a typical fuel cell current density $j = 10$ A.m$^{-2}$.*

- **Solution:** A stationary state can be achieved when $J_w = 0$ in Eq. (10.13). The maximum gradient is therefore

$$\frac{dc_w}{dx} = t_w \frac{j}{F D_w}$$

By introducing the given numbers, we obtain $dc_w/dx = 10^3$ kmol.m$^{-4}$.

## 10.5 Transport of ions and water across ion-exchange membranes

Consider a membrane surrounded by well-stirred, isothermal electrolytes in cells b) or c).

$$\text{Ag(s)} |\text{ AgCl(s)}| \text{ NaCl}, c_1 \parallel \text{NaCl}, c_2 \mid \text{AgCl(s)}|\text{Ag(s)} \qquad \text{(a)}$$

$$\text{H}_2(\text{g}, p) \mid \text{HCl}, c_1 \parallel \text{HCl}, c_2 \mid \text{H}_2 \text{ (g}, p) \qquad \text{(b)}$$

The cells with electrodes are illustrated in Fig. 10.2.

Ions and water can be transported through the membrane. The membrane can sustain a pressure difference. The electrolyte, component 1, is either NaCl (cell a) or HCl (cell b). With only one electrolyte present, the transport coefficients in the membrane can be regarded as constant, and the membrane can be regarded as one interface. The entropy production in the membrane is described by Eq. (10.1). The entropy production in the electrode surfaces was described in the

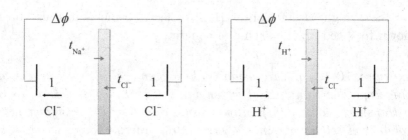

(a) Chloride reversible electrodes     ,  (b) Hydrogen reversible electrodes

Figure 10.2: Schematic illustration of cells a) and b). In cell a) (left) one equivalent of chloride ions is consumed in the electrode reaction on the left-hand side, and is produced on the right-hand side. In cell b), one equivalent of protons is produced in the electrode reaction on the left-hand side and is consumed on the right-hand side. The membrane allows transport of both ions.

most general case by Eq. (9.10). We consider first the isothermal and isobaric membrane, with three driving forces.

### 10.5.1   The isothermal, isobaric system

The three independent forces in the entropy production of the membrane in cells a) and b) are due to a variation across the membrane in concentration, pressure and electric potential. The fluxes of electrolyte, water and charge across the membrane are accordingly

$$J_1 = -\, L_{\mu\mu}\Delta\mu_1 - L_{\mu w}\Delta\mu_w - L_{\mu\phi}\Delta\phi$$
$$J_w = -\, L_{w\mu}\Delta\mu_1 - L_{ww}\Delta\mu_w - L_{w\phi}\Delta\phi$$
$$j = -\, L_{\phi\mu}\Delta\mu_1 - L_{\phi w}\Delta\mu_w - L_{\phi\phi}\Delta\phi \qquad (10.14)$$

The constant temperature in the driving forces has again been included in the coefficients, $L_{ij}$. The last equation gives the electric force, $\Delta\phi$ across the membrane.

$$\Delta\phi = -\frac{L_{1\phi}}{L_{\phi\phi}}\Delta\mu_1 - \frac{L_{w\phi}}{L_{\phi\phi}}\Delta\mu_w - rj \qquad (10.15)$$

where $r = 1/(L_{\phi\phi})$. The first coefficient ratio is the transference coefficient for salt $t_1$, see Eq. (4.51). The second ratio is the transference coefficient for water, $t_w$. Both coefficients are specific for the membrane. The last term is the resistance drop across the membrane.

We use Gibbs-Duhem's equation to express one chemical driving force by the other, $\Delta\mu_w = -(m_1/m_w)\Delta\mu_1$. Here $m_i$ are molalities (in mol.kg$^{-1}$). We approximate $m_1/m_w$ to a constant, and obtain

$$\Delta\phi = - \left( t_1 - t_w \frac{m_1}{m_w} \right) \frac{\Delta\mu_1}{F} - rj \qquad (10.16)$$

At isothermal, isobaric conditions, the contributions from the electrode surfaces to the cell potential cancel. We measure $\Delta\phi$ for a known $\Delta\mu_1$, and compute the value of the parenthesis; which is the apparent transport number, $t'$ [124]. Knowing $t_1$, we can then find $t_w$ from this expression. A finite value of $t_w$ makes $t' \neq t_1$, even if $t_1 = 1$.

The transference coefficient of a salt depends on the choice of electrodes [18]. Consider cell a) with chloride reversible electrodes. From a mass balance in each electrolyte, we have

$$t_1 = t_{M^+} \quad Cl^- \text{-reversible electrodes} \qquad (10.17)$$

Figure 10.2(a) indicates how the electric current is carried by both types of ions, cations and anions. This identification does not depend on the membrane type. In a cation exchange membrane, the transport number of a cation is commonly near unity. Nearly one mole of NaCl is then transferred per mole of electrons passing in the external circuit for such a membrane in cell a). If an anion-exchange membrane is used in cell a), however, a negligible amount of NaCl is transferred between the compartments from left to right. The transference number of the salt in cell a) drops dramatically.

In cell b) where the electrodes are reversible to hydrogen, the transference coefficient of the acid becomes

$$t_1 = -t_{Cl^-} \quad H^+\text{-reversible electrodes} \qquad (10.18)$$

If the membrane is a perfect cation exchange membrane, $t_1 = 0$. If it is a perfect anion-exchanger, $t_1 = -1$. These identifications explain the origin of the electric power in Eq. (10.16). The contribution due to transfer of water is less important, but not negligible.

Transference coefficients have been studied by Okada and coworkers [125, 126, 127, 128] and others [129, 130]. Their studies have shown that ion exchange membranes are seldom perfect. The transport number of the ion in question is typically between 0.98 and 0.90. The water transference coefficient is below 3 for protons, but larger for other ions [128]. In Nafion 115 from Du Pont, $t_w$ was around 10 in the presence of $Na^+$, $K^+$, $Rb^+$ and $Cs^+$, and larger than 10 in the presence of $Ca^{2+}$ and $Sr^{2+}$. It was above 20 in the presence of divalent ions in the cation exchange membrane CR61 from Ionics, varying with membrane composition [18, 126, 127]. The coefficient is negative, when water is transferred in the direction opposite to the electric current. This happens in anion exchange membranes. Values from $-22$ to $-5$ were measured for various anion exchange membranes from Ionics [129]. The absolute value of the water transference number stays high and constant when the electrolyte concentration is not too high ($< 0.1$ kmol.m$^{-3}$), and drops to a low value as the concentration rises above this [18]. Ottøy *et al.* [130] developed a stack method to determine membrane transport numbers of a mixture of ions of varying composition.

Charge transfer and the processes associated with charge transfer are the most important processes in ion-exchange membranes, used for salt power plants, water desalination or fuel cells, see Section 10.6. Diffusion is superimposed on water transfer. This can be described by introducing Eq. (10.14c) into Eqs. (10.14a) and (10.14b):

$$J_1 = -l_{\mu\mu}\Delta\mu_1 - l_{\mu w}\Delta\mu_w + t_1 j/F \qquad (10.19)$$
$$J_w = -l_{w\mu}\Delta\mu_1 - l_{ww}\Delta\mu_w + t_w j/F$$

Electro-osmosis is unavoidable in a system where charge transport takes place. In *reverse electrodialysis* water is forced through a membrane against its chemical potential, leading to the purification of water.

### 10.5.2 The isothermal, non-isobaric system

The transference coefficients $t_1$ and $t_w$ depend on the concentration of electrolyte in the solutions adjacent to the membrane. In order to control this dependence, we study cells with the same concentration on the two sides of the membrane, but with different pressures [131]. The contribution to the streaming potential of the membrane is $(\Delta\phi/\Delta p)_{j=0}$: [131, 129, 18];

$$\left(\frac{\Delta\phi}{\Delta p}\right)_{j=0} = -\frac{1}{F}(t_1 V_1 + t_w V_w) \qquad (10.20)$$

In addition, there is a contribution from the two electrodes. For electrode components in the solid state, there is a contribution equal to

$$\Delta_{\mathrm{el}}\phi_{j=0} = \Delta V_{\mathrm{el}}\Delta p \qquad (10.21)$$

where $\Delta V_{\mathrm{el}} = V_2 - V_1$ is change in reaction volume of the electrodes.

### 10.5.3  The non-isothermal, isobaric system

#### Contributions from the membrane

In the non-isothermal case, there is a heat flux and a thermal driving force in the entropy production of the membrane, Eq. (10.1), in addition to the fluxes and forces in Eqs. (10.14). We obtain

$$J_q'^o = L_{qq}\Delta\left(\frac{1}{T^i}\right) - L_{q\mu}\frac{1}{T^i}\Delta\mu_{1,T} - L_{qw}\frac{1}{T^i}\Delta\mu_{w,T} - L_{q\phi}\frac{1}{T^i}\Delta\phi$$

$$J_1 = L_{\mu q}\Delta\left(\frac{1}{T^i}\right) - L_{\mu\mu}\frac{1}{T^i}\Delta\mu_{1,T} - L_{\mu w}\frac{1}{T^i}\Delta\mu_{w,T} - L_{\mu\phi}\frac{1}{T^i}\Delta\phi$$

$$J_w = L_{wq}\Delta\left(\frac{1}{T^i}\right) - L_{w\mu}\frac{1}{T^i}\Delta\mu_{1,T} - L_{ww}\frac{1}{T^i}\Delta\mu_{w,T} - L_{w\phi}\frac{1}{T^i}\Delta\phi$$

$$j = L_{q\phi}\Delta\left(\frac{1}{T^i}\right) - L_{\phi\mu}\frac{1}{T^i}\Delta\mu_{1,T} - L_{\phi w}\frac{1}{T^i}\Delta\mu_{w,T} - L_{\phi\phi}\frac{1}{T^i}\Delta\phi$$

$$\qquad (10.22)$$

When we use the same solutions, and the same pressure in the two solutions, all terms containing $\Delta\mu_j$ disappear. The definition of the transference coefficients remain the same. There are two driving forces in the entropy production. The corresponding set of flux equations for transport across the membrane is

$$J_q'^o = L_{qq}\Delta\left(\frac{1}{T^i}\right) - L_{q\phi}\frac{1}{T^i}\Delta\phi \tag{10.23}$$

$$j = L_{\phi q}\Delta\left(\frac{1}{T^i}\right) - L_{\phi\phi}\frac{1}{T^i}\Delta\phi \tag{10.24}$$

The coefficient $L_{qq}/(T^iT^o)$ is, as before, identified by the Fourier thermal conductivity of the membrane, and $L_{\phi\phi}/(T^i)$ by the electrical conductivity. The Seebeck coefficient of the membrane is

$$\left[\frac{\Delta\phi}{\Delta T}\right]_{j=0} = -\frac{L_{q\phi}}{T^oL_{\phi\phi}} \tag{10.25}$$

The Peltier heat is defined as

$$\pi^o \equiv \left(\frac{J_q'^o}{j/F}\right)_{\Delta T=0} = F\frac{L_{q\phi}}{L_{\phi\phi}} \tag{10.26}$$

The $\pi^o$ is the reversible heat entering side o at constant $T$, when one mole of electrons are passing the outer circuit from left to right. The Onsager relations can be used to find the relation:

$$\frac{\Delta\phi}{\Delta T} = -\frac{\pi^o}{T^o} \tag{10.27}$$

The reversible heat transferred can be found from the entropy flux, cf. Eq. (3.12), Chapter 3:

$$J_q'^o = T^o\left(J_s - \Sigma_j S_j J_j\right) \tag{10.28}$$

where $S_j$ is the thermodynamic entropy per mol of component $j$. In order to determine the contributions to the Peltier heat, we record all changes in composition in the solution on side o caused by transfer of one mole of electrons. Into this solution, there is a net transport of $t_1$ moles of component 1, accompanied by $t_w$ moles of water. The last term in the equation above, when divided by $j/F$, becomes $\Sigma_j t_j S_j = t_1 S_1 + t_w S_w$. The first term in the parenthesis has contributions from the transported entropy of the ions. If the membrane is a perfect cation-conductor $t_1 = t_{M+} = 1$, we obtain:

$$\frac{\pi^o}{T^o} = \left[S_{M+}^* - (S_1 + t_w S_w)\right] \tag{10.29}$$

The first term in the parenthesis on the right-hand side is due to charge transport in the membrane, Here $S_{M^+}^*$ is the transported entropy of the cation in the membrane.

## Contributions from the electrodes

The two electrodes are kept at different temperatures, and this will also contribute to the emf of the cell. In Chapter 9, we had for reversible electrodes, that $\Delta_{i,o}\phi + \Delta_n G^s = 0$. In the present case, the left (l) and right (r)-hand side chloride reversible electrodes give

$$\Delta_{el,l}\phi = \frac{1}{F}\left[\mu_{Ag}(T) - \mu_{AgCl}(T)\right]$$

$$\Delta_{el,r}\phi = -\frac{1}{F}\left[\mu_{Ag}(T + \Delta T) - \mu_{AgCl}(T + \Delta T)\right] \qquad (10.30)$$

The sum of these contributions is

$$\Delta_{el}\phi = \frac{1}{F}\left[S_{Ag}(T) - S_{AgCl}(T)\right]\Delta T \qquad (10.31)$$

These terms are heat changes associated with the electrode reaction. For cell b) we have, respectively

$$\Delta_{el}\phi = \frac{1}{2}S_{H_2}\Delta T$$

## The Seebeck coefficient of the cell

By combining the contributions from the membrane and the electrodes we obtain the cell potential. Contribution from the transported electron in the connecting leads are usually small. Including these also we obtain the Seebeck coefficient at $j \approx 0$ for cell a):

$$\frac{\Delta\phi}{\Delta T} = \frac{1}{F}\left[-S_{M^+}^* + S_{e^-}^* + (t_w S_w + S_1) + S_{Ag}(T) - S_{AgCl}(T)\right]$$
$$(10.32)$$

The expression on the right-hand side, multiplied with the temperature, is the reversible heat change that is obtained from an entropy balance over the right-hand side electrode compartment, taking into account all components that are transported as a consequence of charge transport. This heat is equal to the heat needed to keep the temperature constant on the right-hand side, during the reversible process.

We see that several large terms contribute to the Seebeck coefficient, unlike in the situation with semiconductors; treated in Chapter 4. This points to an interesting role for ionic conductors as thermoelectric generators [61].

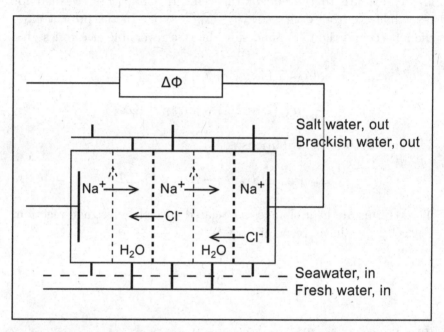

Figure 10.3: The principle of a salt power plant that uses reverse electro-dialysis. The figure shows two unit cells in series. The concentration difference of sodium chloride, between sea water and the fresh water, drives an electric current and creates an electric power.

## 10.6    The salt power plant

In a salt power plant, the energy of mixing of salt water and brackish water is used to give electric power, *cf.* Exercise 2.3.3. The mixing process occurs in a controlled way, with the use of ion exchange membranes. To reach a sizable power, many single units are coupled in series, see Fig. 10.3. The plant operates by reverse electrodialysis.

Sea water and fresh water are fed into alternating compartments, separated by ion-exchange membranes. The membranes alternate between anion- and cation-exchange membranes. One unit cell consists of a salt water compartment, an anion-exchange membrane, a fresh water compartment, a cation-exchange membrane and a salt water compartment. The gradient in chemical potential of salt drives chloride ions through the anion-exchange membrane and sodium ions through the cation-exchange membrane. The transport numbers for the ions are near unity in these membranes. The electric field that arises, causes an electric current in the external circuit. The electrodes of cells a) and b) are now not practical. Cheap, reliable sets are being investigated [132].

Diffusion of salt is prevented by the ion-exchange membranes. Salt may leak into the fresh water compartment and reduce the potential difference. But some salt is also required there to give an acceptable ohmic resistance in this compartment.

To find the total cell potential, we integrate across the cell. The effect of the electrodes is accounted for by the Nernst potential, see Eq. (9.22). When the electrodes are identical and are held at identical conditions, their contributions cancel. We add the contribution from the cation and anion exchange membranes. The first integral goes from high to low salt concentration, while the second integral goes from low to high concentrations. By combining the two integrals, we obtain

$$\Delta\phi_{\text{unit}} = -\left[t_1^C - t_1^A - (t_w^C - t_w^A)\frac{m_1}{m_w}\right]\frac{\Delta\mu_1}{F} - Rj \qquad (10.33)$$

where $Rj$ is the total resistance of the unit cell and the ratio $m_1/m_w$ is the average value for the two salt solutions. A salt concentration ratio of 10:1 at 300 K gives about -110 mV from this formula, for electrodes reversible to the chloride ion. The water transference numbers have opposite signs in the two membranes, so the difference $(t_w^C - t_w^A)$ will be positive, and reduce the effect of the salt gradient. The best performance is obtained with $t_1^C = 1, t_1^A = 0$ and zero or small $t_w^C, t_w^A$.

Increased power can be obtained by stacking unit cells like indicated in the figure. The figure shows two units. The electric potential of the plant is proportional to the number of unit cells. Power densities around 4 $W.m^{-2}$ are now within reach [133].

## 10.7   Concluding remarks

We have seen, through several examples in this chapter, that membranes can be used either to achieve separation or for power generation. There are several options for membrane driving forces; chemical, electrical and thermal. The thermal driving force is so far less explored. This points to a possibility for the use of waste heat sources at low temperatures, adding to *e.g.* the saline power plant.

# Chapter 11

# The state of minimum entropy production

*We present a procedure to minimize the entropy production in a process where defined output conditions are maintained. We introduce optimal control theory as a tool to find this constrained optimum. The procedure is first demonstrated for ideal gas expansion and single plate heat exchange. We show that a process with minimum total entropy production can be characterized by uniform local entropy production for these simple cases. Next, we shall see how optimal control theory can be used in order to find paths of operation with minimum entropy production in two typical process units, the chemical reactor and the distillation tower. Some typical results are presented and concluded with practical guidelines for a second law efficient design.*

It is an aim of the process engineer to reduce the lost work in industrial processes. According to Chapter 2, this means that the total entropy production in the process should be made as small as possible. While we seek to minimize the entropy production, the production in the plant must be maintained, however. This puts at least one constraint on the minimization procedure.

In this Chapter we first discuss entropy production minimization in gas expansion and in heat exchange [134, 135], and demonstrate for these two simple cases, that the path of minimum total entropy production, has constant local entropy production. We shall refer to this result as Equipartion of Entropy Production (EoEP). In these cases we can find analytic results.

Next, we shall show that for more complicated systems, like chemical reactors [49, 136, 137, 138, 139, 140] and distillation columns [141, 142, 143, 144, 145, 146, 147], this equipartition result does not necessarily apply directly, because these units are often too constrained. But we shall see that parts of these units can have (approximately) uniform entropy production. These examples can only be solved with numerical procedures.

This leads us to a hypothesis for the state of minimum entropy production in an optimally controlled system. The hypothesis explains when EoEP is a good approximation to the optimal state [136]. This can then be used for apparatus design in the chemical industry. The knowledge presented in this Chapter shall be used in the end to formulate rules of thumb for some energy efficient process units.

We have chosen chemical reactors and distillation towers as examples because these are central units in chemical process plants. According to Humphrey and Siebert [148], the chemical industry accounts for 27% of the industrial energy demand in the USA, in 1991. The process of distillation was using 40% of that. The chemical reactor has so far not been considered as a work producing or consuming system, as its main purpose is to produce chemicals. With increasing demands on available energy, and boundaries on the earth's ability to deal with entropy production, this may have to change.

The chemical reactor and the distillation towers do not operate isolated, but heavily connected to other process equipments. Minimization of the entropy production in a single process unit alone, may thus lead to increases in other parts of the process. In this Chapter we study the optimal path of the single units. Steps toward a higher energy efficiency can thus only be anticipated when these units can be put in a proper perspective. Systematic efforts in this direction remain to be done, for a start see [149].

We start by explaining the isothermal expansion of an ideal gas, so we can give some exact results. We find the work, the ideal work, the lost work, the entropy production and the minimum entropy production, given certain constraints. We then introduce optimal control theory and explain how this mathematical framework can be used to find the state of minimum entropy production, using the isothermal expansion as an example. We find that the local entropy production is constant throughout the optimal expansion process. This is an example of the theorem of equipartition of entropy production (EoEP) [136, 150, 151, 152, 153, 154]. It says that the entropy production should be constant throughout the process.

## 11.1 Isothermal expansion of an ideal gas

A schematic picture of a container filled with an ideal gas is given in Fig. 11.1. The container is equipped with a piston so that work can be extracted by expanding the gas. Heat is transferred to the gas from the surroundings in order to keep the temperature constant. The temperature of the surroundings and the system is $T_0$ (reversible heat transfer). We consider the expansion of the gas from an initial pressure $p_1$ to a final pressure $p_2$. The corresponding volumes are $V_1$ and $V_2$, respectively. Elementary textbooks are mainly dealing with the reversible version of the process. Only simple irreversible examples, without any constraint on the process duration, are discussed (see for instance [155]). For such simple processes, a complete thermodynamic treatment is possible without introducing any details about time and the dynamics of the process.

Most processes, both in nature and in industry, proceed in a finite period of time, however. We fix the duration of the expansion, $\Theta$, and assume that the movement of the piston can be described by the differential equations

$$\frac{dV(t)}{dt} = -\frac{f}{p(t)^2} \left( p_{\text{ext}}(t) - p(t) \right)$$

$$\Leftrightarrow \quad \frac{dp(t)}{dt} = \frac{f}{N R T_0} \left( p_{\text{ext}}(t) - p(t) \right) \tag{11.1}$$

where $f$ is a constant describing the friction between the piston and the container wall. For the limiting case $f \to \infty$ the system is free of friction. The other symbols are explained in Fig. 11.1. We used

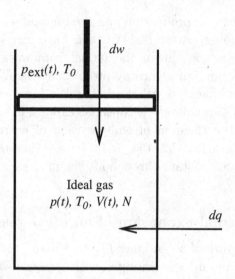

Figure 11.1: A container filled with $N$ mol of an ideal gas with pressure $p(t)$, temperature $T_0$, and volume $V(t)$. The heat $dq$ is added to the gas and work $dw$ is done in a small time interval $dt$. The container is equipped with a piston. The gas expands isothermally against an external pressure $p_{\text{ext}}(t)$. The temperature of the environment is $T_0$.

the ideal gas law to relate $dV(t)/dt$ and $dp(t)/dt$, and we added the factor $1/p(t)^2$ in $dV(t)/dt$ to simplify the mathematical treatment (see Appendix A.5 and Exercise 11.2.1).

## 11.1.1  Expansion work

Expansion produces work. The work that is done *on* the gas is [155]

$$w = -\int_{V_1}^{V_2} p_{\text{ext}}(t)\, dV \qquad (11.2)$$

In the ideal limit, the expansion is a reversible process. In this limit, the external pressure equals the pressure of the gas at all times, and the expansion proceeds infinitely slowly. The ideal work is

$$w_{\text{ideal}} = -\int_{V_1}^{V_2} p\, dV = N\,R\,T_0\,\ln\frac{p_2}{p_1} \qquad (11.3)$$

The work is called ideal since the extracted work $(-w)$ in any other version of the expansion is smaller than $-w_{\text{ideal}}$. An irreversible version of the process has lost work, $w_{\text{lost}} = w - w_{\text{ideal}}$, which is always positive. The name "lost work" reflects that there is potential work which we are not able to extract. The ideal work gives a yardstick, which all other expansions can be compared to. This is the idea behind the second law efficiency (see Chapter 2).

In a $K$-step expansion process against a constant external pressure, the pressure is decreased in $K$ steps to a constant value in each step. The work from Eq. (11.2) becomes

$$w = -N\,R\,T_0 \sum_{i=1}^{K} p_{\text{ext},i} \left( \frac{1}{p_{2,i}} - \frac{1}{p_{1,i}} \right) \tag{11.4}$$

where $p_{\text{ext},i}$, $p_{1,i}$, and $p_{2,i}$ are the external pressure, the initial pressure of the gas and the final pressure of the gas in step number $i$, respectively. Given the values of $p_{\text{ext},i}$ and $p_{1,i}$, one can find $p_{2,i}$ by integrating Eq. (11.1). The corresponding lost work is

$$w_{\text{lost}} = w - w_{\text{ideal}}$$

$$= -N\,R\,T_0 \left[ \sum_{i=1}^{K} p_{\text{ext},i} \left( \frac{1}{p_{2,i}} - \frac{1}{p_{1,i}} \right) + \ln \frac{p_2}{p_1} \right] \tag{11.5}$$

The work in an isothermal expansion (or compression) is often illustrated in a $pV$-diagram. Examples of such diagrams are given in Fig. 11.2. The ideal work of the expansion, Eq. (11.3), is minus the area below the isotherm in these diagrams. The work in a $K = 1$ step expansion, Eq. (11.4), is minus the area of the shaded rectangle in Fig. 11.2(a). The lost work of the same process, Eq. (11.5), is the area between the isotherm and the rectangle in the same figure. Figures 11.2(b)–(d) show expansions with 3, 5 and 15 steps, respectively. We return to this figure when we discuss the entropy production minimization problem later.

## 11.1.2   The entropy production

The entropies of the gas and the surroundings change during the expansion, and the local entropy production is

$$\sigma(t) = \frac{dS^{\text{system}}(t)}{dt} + \frac{dS^{\text{surroundings}}(t)}{dt}$$

$$\sigma(t) = \frac{1}{T_0}\frac{dq_{\text{rev}}}{dt} + \frac{1}{T_0}\left(-\frac{dq}{dt}\right) \tag{11.6}$$

Here, $dq_{\text{rev}}/dt$ is the rate of heat transfer in a reversible expansion between the same initial and final states of the gas. We have taken advantage of entropy being a state function in this calculation. Furthermore, $-dq/dt$ is the rate at which heat is transferred (reversibly) to the surroundings in the irreversible expansion.

Since the expansion is isothermal, the internal energy of the ideal gas, $U$, is constant and the first law gives $dq = -dw$. By using this identity, Eqs. (11.1) and (11.2) in differential form, we can write the local entropy production as

$$\sigma(t) = \frac{1}{T_0}\left(p_{\text{ext}}(t) - p(t)\right)\left(-\frac{dV(t)}{dt}\right)$$

$$= \frac{1}{T_0}\frac{f}{p(t)^2}\left[p_{\text{ext}}(t) - p(t)\right]^2 \tag{11.7}$$

The total entropy production of the expansion is found by integration of the local entropy production over the process duration. With $K \geq 1$ steps, the total entropy production is

$$\frac{dS_{\text{irr}}}{dt} = \sum_{i=1}^{K}\int_{t_{1,i}}^{t_{2,i}}\frac{1}{T_0}\left(p_{\text{ext},i} - p(t)\right)\left(-\frac{dV(t)}{dt}\right)dt$$

$$= -NR\left[\sum_{i=1}^{K}p_{\text{ext},i}\left(\frac{1}{p_{2,i}} - \frac{1}{p_{1,i}}\right) + \ln\frac{p_2}{p_1}\right] \tag{11.8}$$

By comparing this result with the lost work, Eq. (11.5), we see that the Gouy-Stodola theorem, Eq. (2.17), applies. The Gouy-Stodola theorem is generally valid (see e.g. [45] and Section 2.3 for details). In the proof, an important assumption is that all heat is discarded or extracted from a reservoir at the reference temperature $T_0$. We made things simple in this example by assuming that the system and the surroundings are both at $T_0$. Given that they were at another temperature, the Gouy-Stodola theorem would not apply directly. In that case, the Gouy-Stodola theorem applies if an imaginary Carnot machine is used in order to discard or extract the heat from a reservoir at $T_0$ (see Appendix A.6).

### 11.1.3   The optimization idea

We are interested in the path that gives minimum entropy production during expansion from a fixed initial to a fixed final state of the gas. The ideal work is then also fixed, *cf.* Eq. (11.3). Maximizing the work output $(-w)$ and maximizing the second law efficiency are equivalent optimization problems. It makes no sense to, for instance, maximize the work output or minimize the entropy production of this process without fixing the ideal work. Given that the process duration is fixed, maximum work would give an infinite pressure ratio $p_2/p_1$, and minimum entropy production would give $p_2/p_1 = 1$ (no expansion at all).

Without constraints on the duration of the expansion, the minimum entropy production would be a trivial zero, and the maximum work output would be $-w_{\text{ideal}}$, as pointed out above. Since we have a fixed and finite process duration, $\Theta$, the entropy production is not zero and the maximum work output will be lower than $-w_{\text{ideal}}$. For a $K = 1$ step expansion, there is only one external pressure which takes the pressure of the gas from $p_1$ at time 0 to $p_2$ at time $\Theta$. The work and the lost work of this process is illustrated in Fig. 11.2(a). For $K > 1$, there are infinitely many feasible choices of external pressures. This freedom can be used to minimize the entropy production (maximize the work output) of the expansion. The details of this optimization problem are given in Appendix A.5. Figures 11.2(b)–(d) show the work and the lost work for the optimal processes with 3, 5 and 15 steps, respectively.

We are now in a position where we can present the kind of entropy production minimization problems that are central for development of energy efficient processes. By comparing the sub-figures in Fig. 11.2, we see that the work output (the total area of the rectangles)

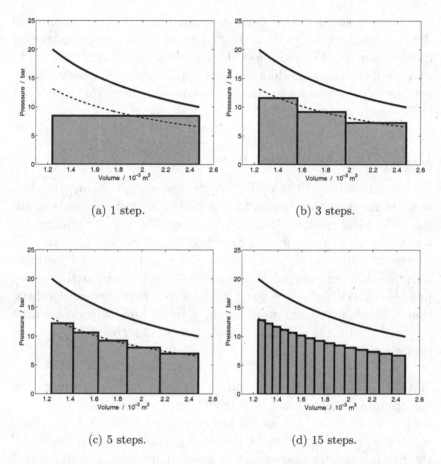

(a) 1 step.                                    (b) 3 steps.

(c) 5 steps.                                   (d) 15 steps.

Figure 11.2: Optimal external pressure vs. volume of the ideal gas. The gray areas are the work output in each expansion step. The lost work is the area between the rectangle(s) and the isotherm (the solid line). The dashed line corresponds to the expansion that produce minimum entropy for the given process duration. ($N = 1$ mol, $T = 298$ K, $p_1 = 20$ bar, $p_2 = 10$ bar, $f = 500 \, \mathrm{m^3 \, Pa/s}$, $\Theta = 10$ s).

increases, and the entropy production decreases, as $K$ increases. Furthermore, the external pressure of the process (the upper edge of the rectangles) approximates the dashed line in the figures better and better, as $K$ increases. The dashed line, characterized by infinitely many steps and a continuously changing external pressure, is a limit for the performance of the process, given that the process duration is fixed.

As a limiting process, it can also be used as a yardstick, much like the reversible process is used when the process is allowed to take infinitely long time. The limiting process in Fig 11.2 is not as general as the reversible limit; it depends on the piston/container used and the dynamics of the system (Eq. (11.1)). In this manner it is a yardstick that can be used in practice to measure the performance [47]. We show in Section 11.2 that the two limiting processes become the same as $\Theta$ goes to infinity.

We say that the system is in the state of minimum entropy production when the expansion proceeds along the dashed line that is given in all the sub-figures of Fig. 11.2. We show in Section 11.2 that this expansion has constant local entropy production throughout. This is one example of the theorem of equipartition of entropy production (EoEP); a result describing the characteristics of the state of minimum entropy production. Later in this Chapter, we shall study the state of minimum entropy production for processes of more practical interest.

## 11.2  Optimal control theory

Consider again the search for the state of minimum entropy production, using the expansion of a piston as example. The dashed line in Fig. 11.2 is the solution of the following optimization problem: Minimize the total entropy production

$$\frac{dS_{\text{irr}}}{dt} = \int_0^\Theta \sigma(t)\, dt = \int_0^\Theta \frac{1}{T_0} \frac{f}{p(t)^2} \left(p_{\text{ext}}(t) - p(t)\right)^2 dt \qquad (11.9)$$

subject to the governing equation for the pressure of the gas

$$\frac{dp(t)}{dt} = \frac{f}{N\,R\,T_0} \left(p_{\text{ext}}(t) - p(t)\right) \qquad (11.10)$$

We have also required that the process duration, $\Theta$, is fixed, and that the initial and final pressures of the gas are $p_1$ and $p_2$, respectively. The optimal profile of the external pressure throughout the process is of interest (dashed line in Fig. 11.2).

This optimization problem can be solved using optimal control theory ([156], see also [157]). In optimal control theory, the variables of the system are divided in two groups; state variables and control variables:

- The state variables are the variables which are governed by differential equations. The pressure of the gas (or alternatively, its volume) is thus a state variable in the present example since it is governed by Eq. (11.10).

- The control variables are our handles on the system, or the means with which we control it. In the present example, the external pressure is the control variable. We assume that we have full control over it, and that it can take any positive value.

Optimal control theory gives a set of necessary conditions for minimum entropy production, much like setting the first derivative equal to zero in ordinary calculus. The first step is to construct the Hamiltonian of the optimal control problem. In our example, the Hamiltonian is

$$\mathcal{H} = \frac{1}{T_0} \frac{f}{p(t)^2} \left( p_{\text{ext}}(t) - p(t) \right)^2 + \lambda(t) \frac{f}{N \, R \, T_0} \left( p_{\text{ext}}(t) - p(t) \right)$$

(11.11)

The Hamiltonian has two parts. The first part is the local entropy production, the integrand in the total entropy production. The second part contains the functional constraints, as products of multiplier functions ($\lambda(t)$'s) and the right hand sides of the governing equations. In this problem, there is only one governing equation, meaning that the second part of the Hamiltonian consists of one term only.

We can benefit from a special property of the Hamiltonian: A general result in optimal control theory is that the Hamiltonian is constant along the coordinate of the system — time, in this case — when it is autonomous [156]. The Hamiltonian is autonomous when it does not depend explicitly on the coordinate of the system (here, time), but only implicitly through the state variables (here, pressure), the control variables (here, external pressure), and the multiplier functions. This is the case for the optimization problems studied in this Chapter.

The necessary conditions for minimum entropy production can thus be derived from the Hamiltonian and consist of differential and algebraic equations. There are two differential equations for each state variable, and one algebraic equation for each control variable. In the present problem the differential equations are

$$\frac{dp(t)}{dt} = \frac{\partial \mathcal{H}}{\partial \lambda} = \frac{f}{NRT_0} \left( p_{\text{ext}}(t) - p(t) \right) \tag{11.12}$$

$$\frac{d\lambda(t)}{dt} = -\frac{\partial \mathcal{H}}{\partial p} = 2 \frac{1}{T_0} \frac{f}{p(t)^2} \left( p_{\text{ext}}(t) - p(t) \right) \frac{p_{\text{ext}}(t)}{p(t)} + \lambda(t) \frac{f}{NRT_0}$$
$$\tag{11.13}$$

In addition, the algebraic equation says that the external pressure should be chosen such that the Hamiltonian is minimized at every instant of time:

$$p_{\text{ext}}(t) = \underset{p_{\text{ext}}(t) > 0}{\text{argmin}} \; \mathcal{H} \tag{11.14}$$

The first differential equation is the governing equation for the pressure, and it is thus not "new". The second differential equation is new and describes the time variation of the multiplier function. In the expansion example, the optimal external pressure is always positive. Equation (11.14) reduces therefore to

$$\frac{\partial \mathcal{H}}{\partial p_{\text{ext}}} = 2 \frac{1}{T_0} \frac{f}{p(t)^2} \left( p(t)_{\text{ext}} - p(t) \right) + \lambda(t) \frac{f}{N \, R \, T_0} = 0 \tag{11.15}$$

The first form of the algebraic equation (11.14) should be used when there are relevant upper and lower bounds on the value of the external pressure.

We have to solve equations equivalent to Eqs. (11.12) to (11.15) in order to find the optimal solution in a given case. Together with appropriate boundary conditions, this is a two point boundary value problem. The problem is usually highly nonlinear and must be solved numerically.

The present problem is simple enough to be solved analytically. We do not use Eq. (11.13), but rather the fact that the Hamiltonian is constant in time. We solve Eq. (11.15) for $\lambda$, introduce the result in the Hamiltonian, and obtain

$$\mathcal{H}^{\min} = \frac{1}{T_0} \frac{f}{p(t)^2} \left( p_{\text{ext}}(t) - p(t) \right)^2 - 2 \frac{1}{T_0} \frac{f}{p(t)^2} \left( p_{\text{ext}}(t) - p(t) \right)^2$$
$$= -\sigma(t) \tag{11.16}$$

*The Hamiltonian reduces to the local entropy production, $\sigma(t)$. The state of minimum entropy production is thus characterized by constant local entropy production, since the Hamiltonian is no function of time. This is an example of the theorem of equipartition of entropy production (EoEP).*

The fact that the local entropy production is constant, can be used to work out all details of the optimal solution analytically, see Exercise 11.2.1. The total entropy production of the optimal solution is

$$\frac{dS_{\mathrm{irr}}}{dt} = \frac{1}{f\, T_0\, \Theta} \left( N\, R\, T_0 \, \ln\left( \frac{p_2}{p_1} \right) \right)^2 \qquad (11.17)$$

This result fits well with intuition. The total entropy production increases when the process duration decreases, when the factor $f$ decreases (more friction), and when $p_2/p_1$ increases. Moreover, the entropy production approaches zero when the process duration goes to infinity, which is in agreement with the reversible process being the ideal limiting process.

Optimal control theory is a general tool to study the state of minimum entropy production and can also be applied, to heat exchange, chemical reactors and distillation columns, for example.

**Exercise 11.2.1** *Start with the fact that the state of minimum entropy production for expansion is characterized by constant local entropy production. Derive how the pressure of the gas and the external pressure vary with time in the optimal state.*

- **Solution:** We express the pressure difference by the local entropy production and obtain

$$(p_{\mathrm{ext}}(t) - p(t)) = -p(t) \sqrt{\frac{T_0\, \sigma_{\mathrm{opt}}}{f}}$$

where $\sigma_{\mathrm{opt}}$ is the constant and optimal entropy production throughout the process. The governing equation for the pressure becomes, from Eq. (11.10),

$$\frac{dp(t)}{dt} = -\frac{1}{N\, R\, T_0} \sqrt{T_0\, \sigma_{\mathrm{opt}}\, f}\; p(t)$$

This is a separable differential equation which can be solved to find the pressure of the gas

$$p(t) = p_1 \exp\left(-\frac{1}{N\,R\,T_0}\,\sqrt{T_0\,\sigma_{\mathrm{opt}}\,f}\,t\right) = p_1\,\left(\frac{p_2}{p_1}\right)^{t/\Theta}$$

and the corresponding external pressure

$$p_{\mathrm{ext}}(t) = p(t)\left(1 - \sqrt{\frac{T_0\,\sigma_{\mathrm{opt}}}{f}}\right)$$

$$= p_1\left(1 + \frac{N\,R\,T_0}{f\,\Theta}\,\ln\left(\frac{p_2}{p_1}\right)\right)\left(\frac{p_2}{p_1}\right)^{t/\Theta}$$

The total entropy production was given in Eq. (11.17). The factor $1/p(t)^2$ in $dV(t)/dt$, *cf.* Eq. (11.1), was introduced in order to simplify the algebra in this exercise and in Appendix A.5.

## 11.3   Heat exchange

Heat exchangers operate with considerable entropy production (*cf.* Exercise 3.3.6 and 3.3.7) and are thus important targets for optimizations. Consider heat exchange across a metal plate that separates a warm and a cold fluid, see Fig. 11.3 [134, 135]. The fluids exchange heat in the $x$-direction. They are transported along the plate in the $z$-direction.

Bejan [45] describes how the various parts of the process, the pump work, the geometry, etc. must be taken into account in a minimization of the entropy production of the total system. We consider only one aspect here, namely the heat exchange, and neglect the other contributions to the entropy production; for instance, pressure drops.

Assume that the hot fluid and the cold fluid are perfectly mixed in the $y$-direction, normal to the paper in Fig. 11.3. Consider a length $\Delta y$ in the $y$-direction. In the $x$-direction, there is a temperature gradient in the metal plate and in the fluid films next to the plate as shown in the lower part of Fig. 11.3. In the bulk of both fluids, there are only gradients in the $z$-direction. Heat is conducted through the metal

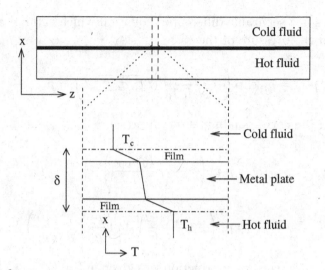

Figure 11.3: Schematic illustration of a heat exchanger. The thermal driving force is taken across $\delta$, the thickness of the metal plate and adjacent film layers.

plate in the $x$-direction. The conduction of heat in other directions is neglected. The system has length $L$, so $0 \leq z \leq L$.

The hot fluid flows from left to right. It enters at temperature $T_{h,\text{in}}$ and leaves at $T_{h,\text{out}}$. The flow rate, $F$, and the heat capacity, $C_p\left(T_h(z)\right)$, are known. The temperature profile of the hot fluid, $T_h(z)$, is then given by conservation of energy:

$$F\, C_p\left(T_h(z)\right)\, dT_h(z) = J_q'(z)\, \Delta y\, dz$$
$$\Rightarrow \quad \frac{dT_h(z)}{dz} = \frac{\Delta y\, J_q'(z)}{F\, C_p\left(T_h(z)\right)} \qquad (11.18)$$

where $J_q'$ is the $z$-dependent measurable heat flux across the metal plate.

None of the properties of the cold fluid (fluid flow, flow direction, heat capacity, etc) are specified yet. We assume that we are able to control the temperature of the cold fluid at every position $z$ in some way. This temperature is our control variable.

## 11.3.1    The entropy production

Heat is transported by conduction through the metal plate and the two adjacent stagnant film layers in the $x$-direction in the system illustrated by Fig. 11.3. The local entropy production is

$$\sigma(x,z) = J_q'(x,z) \frac{d}{dx}\left(\frac{1}{T(x,z)}\right) \qquad (11.19)$$

We deal with stationary states only. In the $x$−direction, this gives

$$\frac{d}{dx}J_q'(x,z) = 0 \quad \Rightarrow \quad J_q'(x,z) = J_q'(z) \qquad (11.20)$$

which means that the measurable heat flux across the metal is independent of $x$. It varies with position $z$, however. We integrate over the $x$-direction and find the entropy production in the plate as a function $z$:

$$\sigma'(z) \equiv \Delta y \int_0^\delta \sigma(x,z)\,dx = \Delta y\, J_q'(z)\left(\frac{1}{T_h(z)} - \frac{1}{T_c(z)}\right)$$
$$\equiv \Delta y\, J_q'(z)\,\Delta\left(\frac{1}{T}\right) \qquad (11.21)$$

where $\sigma'(z)$ denotes the entropy production per unit length in the $z$-direction and where $\delta$ includes the metal plate and its two adjacent stagnant film layers. On the basis of the integrated form, we write the heat flux proportional to its conjugate force

$$J_q'(z) = l_{qq}(z)\,\Delta\left(\frac{1}{T}\right) \qquad (11.22)$$

where the thermal conductivity $l_{qq}(z)$ is the inverse of the integrated thermal resistivity in the $x$−direction:

$$l_{qq}(z) = \left[\int_0^\delta l_{qq}^{-1}(x,z)\,dx\right]^{-1} \qquad (11.23)$$

The coefficient $l_{qq}(z)$ is independent of the thermodynamic force. It can depend on the temperature of either the hot or the cold fluid, but not on both. We take $l_{qq}(z)$ to depend on the temperature of the hot fluid, $T_h(z)$. The total entropy production of the heat exchanger is therefore:

$$\frac{dS_{irr}}{dt} = \Delta y \int_0^L \int_0^\delta \sigma(x, z) \, dx \, dz = \int_0^L \sigma(z)' \, dz$$

$$= \Delta y \int_0^L l_{qq} \left( T_h(z) \right) \left[ \Delta \left( \frac{1}{T} \right) \right]^2 \, dz. \tag{11.24}$$

The entropy production can also be derived from the entropy balance of the hot fluid. The entropy balance has three contributions in addition to the entropy production: entropy of the hot fluid in, $F \, S_{in}$, and out, $F \, S_{out}$ and entropy exchanged with the cold fluid. The entropy balance gives therefore

$$\frac{dS_{irr}}{dt} = F \, S_{out} - F \, S_{in} + \Delta S_c \tag{11.25}$$

where $\Delta S_c$ is the change of the entropy of the cold fluid caused by transfer of heat from it. In a small element $dz$, this change is $-\Delta y \, J_q'(z)/T_c(z) \, dz$. The final expression for the total entropy production, derived from the entropy balance, is therefore

$$\frac{dS_{irr}}{dt} = F \, S_{out} - F \, S_{in} - \Delta y \int_0^L \frac{J_q'(z)}{T_c(z)} \, dz \tag{11.26}$$

The total entropy productions in Eqs. (11.24) and (11.26) are consistent with each other. The proof of this is given as Exercise 11.3.1.

**Exercise 11.3.1** *Show that Eqs. (11.24) and (11.26) are consistent with each other.*

- **Solution:** Equation (11.26) can be rewritten as

$$\frac{dS_{irr}}{dt} = \int_0^L \left[ F \frac{dS_h \left( T_h(z) \right)}{dz} - \Delta y \frac{J_q'(z)}{T_c(z)} \right] dz$$

where we used that the difference between the state functions $S_{out}$ and $S_{in}$ can be calculated using any path. In our model we neglect the effects of pressure drop, so $\frac{dS_h(T_h(z))}{dT_h(z)} = \frac{C_p(T_h(z))}{T_h(z)}$.

By using Eq. (11.18) we obtain

$$
\frac{dS_h\,(T_h(z))}{dz} = \frac{dS_h\,(T_h(z))}{dT_h(z)} \frac{dT_h(z)}{dz}
$$

$$
= \frac{C_p\,(T_h(z))}{T_h(z)} \frac{\Delta y\,J'_q(z)}{F\,C_p\,(T_h(z))} = \frac{\Delta y}{F} \frac{J'_q(z)}{T_h(z)}
$$

which gives

$$
\frac{dS_{\mathrm{irr}}}{dt} = \Delta y \int_0^L J'_q(z) \left( \frac{1}{T_h(z)} - \frac{1}{T_c(z)} \right) dz
$$

$$
= \Delta y \int_0^L l_{qq}\,(T_h(z)) \left[ \Delta \left( \frac{1}{T} \right) \right]^2 dz
$$

This is Eq. (11.24). We have thus two ways to find $dS_{\mathrm{irr}}/dt$; from $l_{qq}$, $T_h(z)$ and $T_c(z)$ and from Eq. (11.26). This is an example of how the consistency of a model can be verified, as discussed in Section 2.5.

## 11.3.2  Optimal control theory and heat exchange

Optimal control theory can be applied to find the properties of the state of minimum entropy production in heat exchange. We want to minimize the total entropy production, Eq. (11.24), subject to Eq. (11.18) at every position $z$. In addition, we fix the total heat transported across the metal plate by fixing $T_{h,\mathrm{in}}$ and $T_{h,\mathrm{out}}$. The Hamiltonian of the optimal control problem is

$$
\mathcal{H} = \Delta y\, l_{qq}\,(T_h(z)) \left[ \Delta \left( \frac{1}{T} \right) \right]^2 + \lambda(z) \frac{\Delta y\,J'_q(z)}{F\,C_p\,(T_h(z))} = \text{constant}
$$

(11.27)

where $\lambda(z)$ is a multiplier function. The first term in the Hamiltonian is the local entropy production in Eq. (11.24) and the second term is the product of the multiplier function and the right hand side of Eq. (11.18). The Hamiltonian is constant, because it does not depend explicitly on $z$.

The necessary conditions for minimum entropy production are then the differential equations

$$
\frac{dT_h(z)}{dz} = \frac{\partial \mathcal{H}}{\partial \lambda} \qquad \frac{d\lambda(z)}{dz} = -\frac{\partial \mathcal{H}}{\partial T_h} \qquad (11.28)
$$

and the algebraic equation

$$\frac{\partial \mathcal{H}}{\partial T_c} = \Delta y \, l_{qq} \left( T_h(z) \right) \left[ 2 \Delta \left( \frac{1}{T} \right) + \frac{\lambda(z)}{F \, C_p \left( T_h(z) \right)} \right] \frac{1}{T_c(x)^2} = 0$$

$$(11.29)$$

The left part of Eq (11.28) reduces directly to Eq. (11.18), whereas the right part is a new differential equation which gives the position dependence of the multiplier function. The boundary conditions for these differential equations are the fixed values of $T_{h,\text{in}}$ and $T_{h,\text{out}}$. We have only given the weak form of the necessary condition for $T_c(z)$, Eq. (11.29), because we assume that the cold fluid can have any temperature at every position $z$. Optimal control theory gives a stronger condition when there are significant upper and/or lower bounds on $T_c(z)$ (*cf.* Section 11.1), and is well suited to handle more complicated cases than here.

We solve the optimization problem in the same way as we solved the expansion problem in Section 11.1. The new differential equation for the multiplier function is not used. Instead, we solve Eq. (11.29) for $\lambda(z)$ and introduce the result in the Hamiltonian, Eq. (11.27). After some rearrangements we obtain

$$\mathcal{H} = \Delta y \, l_{qq} \left( T_h(z) \right) \left[ \Delta \left( \frac{1}{T} \right) \right]^2 - 2 \, \Delta y \, J_q'(z) \, \Delta \left( \frac{1}{T} \right)$$

$$= - \sigma(z) = \text{constant} \qquad (11.30)$$

This means that the state of minimum entropy production is again characterized by constant local entropy production (EoEP).

An interesting question is whether equipartition of the thermal driving force (EoF) gives a good approximation to the real optimum. The two equipartition results are trivially equivalent when $l_{qq}(z)$ is constant. When this is not the case, EoEP always gives the lower total entropy production, but the difference is negligible if the heat transfer duty is reasonably high [135]. This is illustrated in Exercise 11.3.2.

**Exercise 11.3.2** *A hot stream needs to be cooled from 450 K ($T_{h,in}$) to 400 K ($T_{h,out}$). The heat exchanger has a heat transfer coefficient $l_{qq} = U \, T_h(z)^2$, and the heat capacity of the hot stream is taken to be constant.*

a) Derive the temperature profile, the thermal driving force and the local and total entropy production for the EoEP case.

b) Repeat the derivations in a) for the EoF case.

c) Use $U = 340\ W.K^{-1}.m^{-2}$, $\Delta y = 1\ m$, $L = 10\ m$ and $F\,C_p = 1200\ J.K^{-1}$. Compare the total entropy productions of the EoEP and EoF cases.

**Solution:**

a) From Eq. (11.21) it follows for the EoEP case that

$$\sigma_{\mathrm{EoEP}} = \Delta y\, U\, T_h(z)^2 \left( \Delta \left( \frac{1}{T} \right) \right)^2$$

$$\Rightarrow \quad \Delta \left( \frac{1}{T} \right) = - \sqrt{\frac{\sigma_{\mathrm{EoEP}}}{\Delta y\, U}}\, \frac{1}{T_h(z)}$$

By introducing this in Eq. (11.18) we find

$$\frac{1}{T_h(z)} \frac{dT_h(z)}{dz} = \frac{d \ln T_h(z)}{dz} \quad \text{is constant}$$

$$\Rightarrow \quad T_h(z) = T_{h,\mathrm{in}} \left( \frac{T_{h,\mathrm{out}}}{T_{h,\mathrm{in}}} \right)^{z/L}$$

We then find the thermal driving force by rearranging Eq. (11.18) and introducing this temperature profile. The thermal driving force is

$$\Delta \left( \frac{1}{T} \right)_{\mathrm{EoEP}} = \frac{F\,C_p}{\Delta y\, U} \frac{1}{T_h(z)^2} \frac{dT_h(z)}{dz}$$

$$= \frac{F\,C_p}{\Delta y\, U\, L} \ln \left( \frac{T_{h,\mathrm{out}}}{T_{h,\mathrm{in}}} \right) \frac{1}{T_h(z)}$$

The resulting local entropy production is

$$\sigma_{\mathrm{EoEP}} = \frac{1}{\Delta y\, U} \left[ \frac{F\,C_p}{L} \ln \left( \frac{T_{h,\mathrm{out}}}{T_{h,\mathrm{in}}} \right) \right]^2$$

and the total entropy production is

$$\left( \frac{dS_{\mathrm{irr}}}{dt} \right)_{\mathrm{EoEP}} = \frac{1}{\Delta y\, L\, U} \left[ F\,C_p \ln \left( \frac{T_{h,\mathrm{out}}}{T_{h,\mathrm{in}}} \right) \right]^2 .$$

**b)** When the thermal driving force is constant, we find from Eq. (11.18) that

$$\frac{1}{T_h(z)^2}\frac{dT_h(z)}{dz} \quad \text{is constant}$$

$$\Rightarrow \quad \frac{1}{T_h(z)} = \frac{1}{T_{h,\text{in}}} + \left(\frac{1}{T_{h,\text{out}}} - \frac{1}{T_{h,\text{in}}}\right)\frac{z}{L}$$

By rearranging Eq. (11.18) and introducing this temperature profile, we find the thermal driving force:

$$\Delta\left(\frac{1}{T}\right)_{\text{EoF}} = \frac{F\,C_p}{\Delta y\,U}\frac{1}{T_h(z)^2}\frac{dT_h(z)}{dz}$$

$$= -\frac{F\,C_p}{\Delta y\,U\,L}\left(\frac{1}{T_{h,\text{out}}} - \frac{1}{T_{h,\text{in}}}\right)$$

which means that the local and the total entropy productions are

$$\sigma_{\text{EoF}} = \frac{1}{\Delta y\,U}\left[\frac{F\,C_p}{L}\left(\frac{1}{T_{h,\text{out}}} - \frac{1}{T_{h,\text{in}}}\right)\right]^2 T_h(z)^2$$

and

$$\left(\frac{dS_{\text{irr}}}{dt}\right)_{\text{EoF}}$$

$$= \frac{1}{\Delta y\,U}\left[\frac{F\,C_p}{L}\left(\frac{1}{T_{h,\text{out}}} - \frac{1}{T_{h,\text{in}}}\right)\right]^2 \int_0^L T_h(z)^2\,dz$$

$$= \frac{1}{\Delta y\,L\,U}\left[F\,C_p\left(\frac{1}{T_{h,\text{out}}} - \frac{1}{T_{h,\text{in}}}\right)\right]^2 T_{h,\text{out}}\,T_{h,\text{in}},$$

respectively.

**c)** When we introduce the numerical values the parameters, we find $(dS_{\text{irr}}/dt)_{\text{EoEP}} = 5.8756$ J.K$^{-1}$.s$^{-1}$ and $(dS_{\text{irr}}/dt)_{\text{EoF}} = 5.8824$ J.K$^{-1}$.s$^{-1}$. The EoF-value is indeed larger than the EoEP-value, which is the real optimum solution. The difference is only 1% and one may therefore conclude also that the use of the theorem of equipartition of the thermodynamic force leads to a very good approximation of the state of minimum entropy production. The temperature profiles of the EoEP- and EoF-solutions are shown in Fig. 11.4.

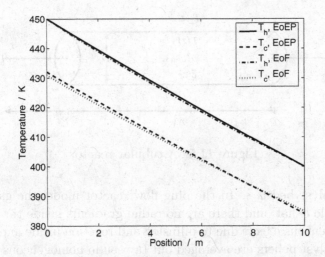

Figure 11.4: The temperature profiles of the EoEP- and EoF-solutions in Exercise 11.3.2c.

The temperature profiles of the hot and the cold fluid are approximately parallel when the production of entropy is minimum (see Fig. 11.4). How can this be achieved in a real heat exchanger? The heat capacity of the cold fluid is a central variable in this context. The heat capacity may be included as a variable in the optimization, and matched with a medium afterward. Full agreement with the optimum can be impractical or not economically feasible. It is therefore good to be able to approximate optimal heat exchange conditions.

For simple heat exchange processes, i.e. processes without phase change and small/moderate temperature change in each stream, counter-current heat exchange with a properly adjusted flow rate of the utility is probably the best approximation. For more complex situations, a series of counter-current, and/or cross-current heat exchangers might be needed to achieve a good approximation.

## 11.4   The plug flow reactor

Consider a tubular reactor with diameter $D$ and length $L$ as sketched in Fig. 11.5. The reactor is filled with catalyst pellets with diameter $D_p$ and density (per unit volume of reactor) $\rho_B$. The void fraction of

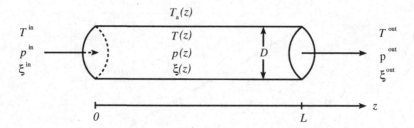

Figure 11.5: A tubular reactor.

the catalyst bed is $\epsilon$. In the plug flow reactor model the gas velocity profile is flat, and there are no radial gradients inside the reactor. Heterogeneous effects due to diffusion and reaction inside and around the catalyst pellets are averaged out (a pseudo homogeneous model). Transport in the $z$-direction is only by convection. A cooling/heating medium with temperature $T_a(z)$ is placed on the outside of the reactor wall in order to remove/supply heat.

We consider a mixture of reacting gases with $n$ components. At the catalyst surface $m$ reactions take place. The reacting mixture is specified by state variables; temperature $T(z)$, pressure $p(z)$ and degrees of reactions $(\xi_j(z), j = 1, ..., m)$. The variables are governed by the balance equation for energy, momentum (Ergun's equation) and mass, see Table 11.1.

We shall find the $T_a$-profile that minimizes the entropy production of the reactor for a given chemical conversion. In other words, we are looking for the optimal heating/cooling strategy of the reactor, with $T_a(z)$ as the control variable.

## 11.4.1   The entropy production

The entropy production of the reactor can be found in two ways. At stationary states, the entropy balance has three contributions in addition to the entropy production. Entropy follows the flows in, $(\sum_i F_i S_i)_{\text{in}}$, and out, $(\sum_i F_i S_i)_{\text{out}}$. In addition, heat is exchanged with the utility. The entropy balance gives therefore

$$\frac{dS_{\text{irr}}}{dt} = \left(\sum_i F_i S_i\right)_{\text{out}} - \left(\sum_i F_i S_i\right)_{\text{in}} + \Delta S_{\text{u}} \qquad (11.31)$$

Table 11.1: Governing equations for the stationary state plug flow reactor.

---

**Balance equation for internal energy:**

$$\frac{dT}{dz} = f_T = \frac{\pi D J_q' + \Omega \rho_B \sum_j [r_j (-\Delta_r H_j)]}{\sum_i [F_i C_{p,i}]}$$

---

**Momentum balance (Ergun's equation):**

$$\frac{dp}{dz} = f_p = -\left( \frac{150 \mu}{D_p^2} \frac{(1-\epsilon)^2}{\epsilon^3} + \frac{1.75 \rho^0 v^0}{D_p} \frac{1-\epsilon}{\epsilon^3} \right) v$$

---

**Mole balances:**

$$\frac{d\xi_j}{dz} = f_{\xi_j} = \frac{\Omega \rho_B}{F_A^0} r_j \qquad j = 1, ..., m$$

---

**Balance equation for entropy:**

$$\frac{dS_{irr}}{dt} = S_{out} - S_{in} - \pi D \int_0^L \frac{J_q'}{T_a} dz$$

$$= \int_0^L \left[ \Omega \rho_B \sum_j \left[ r_j \left( -\frac{\Delta_r G_j}{T} \right) \right] + \pi D J_q' \Delta \frac{1}{T} - \Omega \frac{v}{T} \frac{dp}{dz} \right] dz$$

---

where $\Delta S_u$ is the change of the entropy of the utility caused by transfer of heat to or from it. In a small element $dz$, this change is $-\pi D J_q'(z)/T_a(z) \, dz$, where $J_q'(z)$ is the heat flux from the utility to the reacting stream at $z$. The final expression for the total entropy production, derived from the entropy balance, is therefore

$$\frac{dS_{irr}}{dt} = \left( \sum_i F_i S_i \right)_{out} - \left( \sum_i F_i S_i \right)_{in} - \pi D \int_0^L \frac{J_q'(z)}{T_a(z)} dz \quad (11.32)$$

The entropy production can also be calculated from the fluxes and forces. We saw in Chapter 6 that there are three phenomena that produce entropy in a plug flow reactor: Reactions, heat transport through the reactor wall and frictional flow (pressure drop). The local entropy production (on a unit length basis) was from

Eq. (6.34)

$$\sigma' = \Omega\,\rho_{\mathrm{B}} \sum_{j}\left[ r_j \left( -\frac{\Delta_{\mathrm{r}}G_j}{T} \right) \right] + \pi\,D\,J_q'\,\Delta\frac{1}{T} + \Omega\,v\left( -\frac{1}{T}\frac{dp}{dz} \right)$$

(11.33)

Each term in Eq. (11.33) is a product of a flux and its conjugate force. The first term is sum over all reactions; the flux is the reaction rate, $r_j$, and the chemical force is $-\Delta_{\mathrm{r}}G_j/T$. This term was discussed in depth in Chapter 7.

The second term is due to heat transfer; the flux is the heat flux, $J_q'$, and the thermal force is $\Delta 1/T = 1/T - 1/T_{\mathrm{a}}$. The last term is due to frictional flow; the flux is the gas velocity, $v$, and the force is $(-1/T)\,(dp/dz)$.

The total entropy production is the integral of $\sigma$ over the reactor coordinate $z$

$$\frac{dS_{\mathrm{irr}}}{dt} = \int_0^L \sigma'\,dz$$

(11.34)

The two expressions for the total entropy production, Eqs. (11.32) and (11.34), are equivalent. The proof of this is given in Exercise 11.4.1. We continue to use Eq. (11.34), because it enables us to connect with local variables.

It can be shown that the reactor produces or consumes work in the same way as the heat exchanger, see Appendix A.6. More precisely, an endothermic reactor is a work consuming apparatus and an exothermic reactor is a work producing apparatus. This means that minimizing the entropy production and minimizing the work requirement of the reactor are equivalent problems when the state of the reacting mixture is fixed at the inlet and at the outlet.

**Exercise 11.4.1** *Show that Eqs. (11.32) and (11.34) are equivalent. Assume that the reacting stream is a mixture of ideal gases.*

**Solution:** The starting point is the total entropy balance Eq. (11.32), which we rewrite as:

$$\frac{dS_{\mathrm{irr}}}{dt} = \int_0^L \left( \frac{d\sum_i F_i\,S_i}{dz} - \pi\,D\,\frac{J_q'}{T_{\mathrm{a}}} \right) dz$$

from which we recognize the local entropy production as

$$\sigma' = \frac{d \sum_i F_i \, S_i}{dz} - \pi \, D \, \frac{J_q'}{T_a}$$

$$= \sum_i \left[ F_i \left( \frac{\partial S}{\partial T} \right)_{p,F_i} \frac{dT}{dz} + F_i \left( \frac{\partial S}{\partial p} \right)_{T,F_i} \frac{dp}{dz} + \left( \frac{\partial F_i \, S_i}{\partial F_i} \right)_{T,p} \frac{dF_i}{dz} \right]$$

$$- \pi \, D \, \frac{J_q'}{T_a}$$

with one parenthesis $[]_i$ for each component. The temperature, the pressure and the flow rates describe the reacting mixture. In order to derive the local entropy production we need the governing equations (Table 11.1) and the derivatives of the entropy with respect to temperature, pressure and flow rates. We use another form of the mole balances here:

$$\frac{dF_i}{dz} = \Omega \, \rho_B \sum_{j=1}^m \nu_{j,i} r_j \qquad i = 1, ..., n$$

The partial molar entropy of component $i$ for an ideal gas is:

$$S_i = S_i^\ominus - R \, \ln \frac{p \, x_i}{p^\ominus}$$

The derivatives of the entropy with respect to temperature, pressure and flow rates become:

$$\sum_i F_i \left( \frac{\partial S_i}{\partial T} \right)_{p,F_i} = \frac{1}{T} \sum_i [F_i \, C_{p,i}],$$

$$\sum_i F_i \left( \frac{\partial S}{\partial p} \right)_{T,F_i} = -\frac{R}{p} \sum_i F_i = -\frac{\Omega \, v}{T},$$

$$\left( \frac{\partial \sum_i F_i \, S_i}{\partial x_i} \right)_{T,p} = S_i = S_i^\ominus - R \, \ln \frac{p \, x_i}{p^\ominus},$$

By introducing the balance equations and the derivatives of the entropy in the local entropy production, we obtain

$$\sigma = \pi D \frac{J'_q}{T} + \Omega \, \rho_B \sum_{j=1}^{m} \left[ r_j \, \frac{\Delta_r H_j}{T} \right] + \Omega \, v \left( -\frac{1}{T} \frac{dp}{dz} \right)$$

$$+ \Omega \, \rho_B \sum_{i=1}^{n} \sum_{j=1}^{m} [\nu_{j,i} \, r_j \, S_i] - \pi D \frac{J'_q}{T_a}$$

$$= \Omega \, \rho_B \sum_{j=1}^{m} \left[ r_j \left( -\frac{\Delta_r G_j}{T} \right) \right] + \pi D \, J'_q \, \Delta \frac{1}{T} + \Omega \, v \left( -\frac{1}{T} \frac{dp}{dz} \right)$$

which is the local entropy production as given in Eq. (11.33). In the last equality, we used $\sum_{i=1}^{n} \sum_{j=1}^{m} [\nu_{j,i} \, r_j \, S_i] = \sum_{j=1}^{m} [r_j \, \Delta_r S_j]$. The derivation is similar for a nonideal gas mixture. We then have to change the balance equation for the internal energy to

$$C_p \frac{dT}{dz} = \pi D \, J'_q + \Omega \, \rho_B \sum_{j} [\, r_j \, (-\Delta_r H_j)] - \left( \frac{\partial H}{\partial p} \right)_{T, F_i} \frac{dp}{dz}$$

in order to account for the change of enthalpy caused by change of pressure.

## 11.4.2    Optimal control theory and plug flow reactors

We find necessary conditions for minimum entropy production using optimal control theory [156] in the same way as for the expansion and the heat exchanger, see Sections 11.1 and 11.3. The Hamiltonian for the present problem is

$$\mathcal{H} = \sigma + \lambda_T \, f_T + \lambda_p \, f_p + \sum_{j} [\, \lambda_{\xi_j} \, f_{\xi_j}] \qquad (11.35)$$

The Hamiltonian contains the local entropy production Eq. (11.33), and products of multiplier functions, $\lambda$'s, and the right-hand sides of the conservation equations in Table 11.1. The Hamiltonian does not depend explicitly on $z$ and is thus constant along the $z$-coordinate in the state of minimum entropy production [156].

The necessary conditions for minimum entropy production are the following $2m + 4$ differential equations

$$\frac{dT}{dz} = \frac{\partial \mathcal{H}}{\partial \lambda_T} \qquad\qquad \frac{d\lambda_T}{dz} = -\frac{\partial \mathcal{H}}{\partial T}$$

$$\frac{dp}{dz} = \frac{\partial \mathcal{H}}{\partial \lambda_P} \qquad\qquad \frac{d\lambda_p}{dz} = -\frac{\partial \mathcal{H}}{\partial p}$$

$$\frac{d\xi_j}{dz} = \frac{\partial \mathcal{H}}{\partial \lambda_{\xi_j}} \qquad\qquad \frac{d\lambda_{\xi_j}}{dz} = -\frac{\partial \mathcal{H}}{\partial \xi_j} \qquad (11.36)$$

where $j = 1, ..., m$, and the algebraic equation

$$\frac{\partial \mathcal{H}}{\partial T_a} = 0 \qquad \text{for all } z \in [0, L] \qquad (11.37)$$

The left column in Eq. (11.36) reduces to the balance equations, see Table 11.1.

It is impossible to eliminate all the multiplier functions from the Hamiltonian in the same way as we did in Sections 11.2 and 11.3.2. The reason is that there are less control variables than there are thermodynamic forces in the system (one control variable and $m + 2$ forces). This property of the problem also has consequences for the validity of the equipartition results (EoEP and EoF). As shown in Appendix A.7, EoEP and EoF can be proven rigorously only when the number of control variables is equal to or larger than the number of thermodynamic forces in the system [136]. This means that EoEP and EoF do not describe the state of minimum entropy production in the entire reactor. Equipartition may describe the optimal state in parts of the reactor, though.

### 11.4.3   A highway in state space

The necessary conditions, Eqs. (11.36)–(11.37), have to be solved numerically. Boundary conditions are needed together with the differential equations in Eq. (11.36). For a fixed chemical conversion ($\xi_j$ fixed at the inlet and the outlet), the optimal state depends on the temperatures and the pressures at the inlet and the outlet. By studying all possible combinations of temperatures and pressures, an enormous collection of optimal states is obtained. The collection becomes even larger when we study the effect of varying the composition at both boundaries. We focus on a set of solutions crowding in on what seems to be a *highway in state space* [136]. This is a general

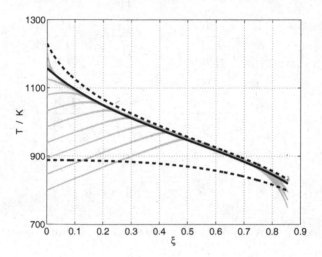

Figure 11.6: State diagram for minimum entropy production for the $SO_2$ oxidation reactor. Printed with permission of Chemical Engineering Science.

property of the states of minimum entropy production in plug flow reactors. We use the oxidation of $SO_2$ to $SO_3$ as an example reaction, and neglect pressure gradients. The reaction is exothermic. Thermodynamic and kinetic data and other model parameters for the reactor can be found in Ref. [49].

The state of the reaction mixture in the $SO_2$ oxidation reactor is specified by the degree of reaction and the temperature of the mixture. Figure 11.6, which gives $T(z)$ as a function of $\xi(z)$, is a dynamic state diagram for the reactor. There are several lines in Fig. 11.6. The upper dashed line is the equilibrium line. Along this line, $\Delta_r G$ and the reaction rate are both zero. The lower dashed line is the maximum reaction rate line. For any value of the conversion, the temperature given by this line corresponds to maximum reaction rate. Furthermore, we have plotted the optimal solutions for fifty combinations of ten inlet temperatures and five outlet temperatures (gray solid lines). Only ten gray lines can be distinguished in the figure. The solutions reveal an interesting property: The central parts of the solutions fall more or less on the same line. This line extends from the inlet on the left-hand side to the outlet on the right-hand side. The individual solutions enter and leave this line at different positions depending on where their initial and final destinations are. The collection of solutions in Fig. 11.6 looks like a highway with its connecting roads. Johannessen and Kjelstrup [136] adopted the highway picture and called the band, which all solutions enter, a "reaction highway".

The highway coincides with the black solid line in Fig. 11.6. This is the solution of the same optimization problem when there is no resistance to heat transfer, when $U = \infty$. The reaction is then the only entropy producing process in the reactor, because we neglected the pressure drop here. The position of the pure reaction solution in state space, and thus the highway, is dictated by the process intensity (chemical conversion per meter reactor). For an infinitely long reactor, the highway coincides with the equilibrium line. As the process intensity increases, the highway shifts toward the maximum reaction rate line. The highway in Fig. 11.6 corresponds to a process intensity of technical relevance. Johannessen and Kjelstrup [49] showed that there exists an optimal reactor length which gives the best trade off between low entropy production of heat transfer and reactions (long reactors are favorable) and low entropy production of frictional flow or pressure drop (short reactors are favorable).

The nature of the highway is further presented in Fig. 11.7. The figure shows the local entropy production as a function of the conversion for one solution (solid line). The part of the solution with approximately constant local entropy production is on the highway. The contributions from the reaction (dashed line) and the heat transfer term (dash-dot line) are also included. The entropy production is not constant at the beginning and the end of the reactor because the reactor has to accommodate certain boundary conditions. These parts of the solution are off the highway. The figure shows that there is a shift of operation mode as the solution enters the highway. Up to this point the entropy production due to the reaction is much larger than the heat transfer term. We say that the reactor operates in a *reaction mode*. The reactor operates with low heat transfer duty in this region. It is basically the enthalpy of reaction that heats the reaction mixture until it reaches the highway. Once on the highway, the heat transfer term dominates the entropy production. It is the heat transfer that drives the solution along the highway, and we can say that the reactor is in a *heat transfer mode* of operation. There

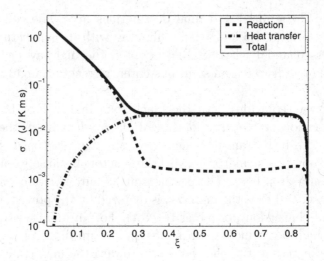

Figure 11.7: Local entropy production as a function of conversion for one solution of the optimization problem when we neglect the pressure drop. Printed with permission of Chemical Engineering Science (publication CES5877).

is a fine balance between the heat produced by the reaction and the rate of heat transfer. This fine balance prevents the reaction from reaching equilibrium and is therefore essential for the chemical production on the highway.

There is also a highway in state space when we include the pressure drop in the optimization. It exists for other reactors too, with one or more reactions which might be exothermic or endothermic. When the process intensity is not too high, EoEP and EoF are good approximations to the optimal states on the highway. EoEP is slightly better than EoF. For high process intensities, the highway is shifted far away from the equilibrium line. The nonlinearities in the reaction rate expressions are then substantial, and EoEP and EoF do not approximate the highway any more. The process intensity in typical industrial reactors is usually in the range where EoEP and EoF approximate the highway well.

It is interesting that the highway can be characterized by almost constant entropy production and forces. As shown in Appendix A.7, these properties have been proven when there are enough control variables, and the flux-force relations are linear [136, 151, 152]. But they are more general than that: the system adjusts to some kind of optimal behavior, if it is given enough freedom to do so, as well as when the flux is a nonlinear function of the force as in chemical reactors. It is clear that there is enough freedom for adjustments in the central part of the reactors. These findings were summarized as a hypothesis for the state of minimum entropy production in an optimally-controlled system by Johannessen and Kjelstrup [136]:

> *EoEP, but also EoF are good approximations to the state of minimum entropy production in the parts of an optimally controlled system that have sufficient freedom to equilibrate internally.*

The concept of "sufficient freedom" needs some explanation. A system where the number of control variables is at least as high as the number of forces has sufficient freedom throughout to equilibrate internally — provided that it is not too far from equilibrium. A system with too few control variables has generally not enough freedom in the whole system. Boundary conditions as well as the compromise that must take place between the dissipative phenomena, will restrict the solution. The central part of the system is relatively more free from these restrictions. Freedom is thus not only related to the number of control variables, but also to the number and type of constraints on the system. The sufficient freedom is then necessarily system specific.

### 11.4.4 Reactor design

The discussions in the previous sections, about EoF, EoEP and highways in state space, indicate that solutions with minimum entropy productions have some common properties. These properties can now be exploited to give guidance to the designing of process equipment which lead to energy-efficient operation. We know that the contribution to the entropy production from heat transfer is often the largest source of dissipation; in chemical reactors, heat exchangers and distillation columns. The first step in a strategy to increase the energy efficiency in these systems, should thus be to make the heat transfer as efficient as possible. This can be linked to the statement made by Leites *et al.* [159] in their first commandment: the driving force of a process must approach zero at all points in a reactor, at all times. A

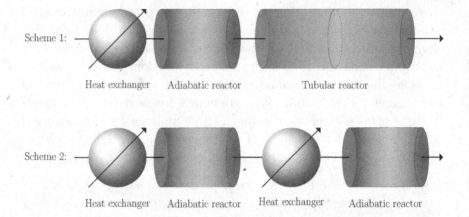

Figure 11.8: Process configurations for an energy-efficient reactor design according to Wilhelmsen *et al.* [158]. Reprinted with permission from The Royal Society of Chemistry.

thermal driving force can be made small by increasing the heat transfer coefficients or the surface area. The interesting question beyond that becomes: What can be done, once the heat transfer has been made as efficient as possible?

In all optimal reactor solutions presented in the literature [49, 136, 137, 138, 139, 140], the optimal solutions first enter a reaction mode at the inlet, before it proceeds into a heat transfer mode of operation in the central part (see Fig. 11.7).

It follows for single tubular reactors of length $L$, that a (close to) adiabatic inlet section, $L_1$, is an advantage for the total entropy production. Furthermore, the next part, $L_2$, can be best characterized by equipartition of the entropy production and, in some cases, also by the equipartition of the forces. In other words, finding the optimal solution for a system, translates into a procedure where one considers a scheme with separate units, like that illustrated in Fig. 11.8 [140, 158]. The reactor part of the system consists of two subunits, an adiabatic pre-reactor and a tubular reactor with heat transfer. To complete the system analysis, a heat exchanger is added in front of the adiabatic reactor as in Scheme 1. This system can now be used to account for the trade-off between the contributions to the entropy production, including the contribution to the entropy production from the heat exchange upfront of the reactor system.

The purpose of the heat exchanger (the first item in Scheme 1) is to bring the reacting mixture to the optimal initial temperature. The purpose of the adiabatic reactor, the next unit, is to operate the chemical reactor in reaction mode. Whether it pays, in terms of entropy production, to use Scheme 1 with the reactor in the heat-transfer mode of operation, or to transfer to more discrete units, as illustrated in Scheme 2 in Fig. 11.8, depends on the relative values of the heat transfer coefficients. When the heat-transfer coefficients across the reactor tube wall are very low, it is better to use dedicated heat exchangers for heat transfer. It will then be beneficial to split the operation in heat-transfer mode (taken care of by the tubular reactor in Scheme 1) by separate sets of one or more adiabatic reactor stages with interstage heating/cooling, as illustrated in Scheme 2 of Fig. 11.8. Two or more heat exchanger-adiabatic reactor pairs may be cost-effective and energy-efficient as well. This shows that a complex optimal control problem can be reduced, if not avoided, and that the process of finding an energy-efficient reactor design can be simplified significantly. As an example, de Koeijer *et al.* [160] found the second-law optimal path of a four-bed $SO_2$ converter with intercoolers.

## 11.5   Distillation columns

Mixtures of chemicals are often separated by distillation. Distillation requires addition and withdrawal of heat. This is connected with a large entropy production. Distillation columns are therefore targets for optimization.

A sketch of a tray distillation column is shown in Fig. 11.9. The column has $N$ trays which bring a rising stream of vapor in close contact with a descending stream of liquid. The feed stream, $F$, is separated into two fractions with different boiling temperatures and compositions. The low-boiling fraction is called the distillate, $D$, and is taken out at the top of the column. The high-boiling fraction is obtained as the bottom stream, $B$. In a conventional column, heat is only added in a reboiler and only withdrawn in a condenser, $Q_{N+1}$

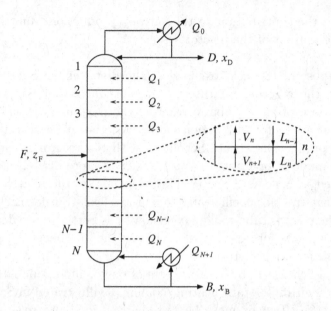

Figure 11.9: A tray distillation column with heat exchange on all trays. $Q_1,...,Q_N$ are zero in an adiabatic column. All $Q$'s are in general nonzero in a diabatic column.

and $Q_0$ in Fig. 11.9. No heat is added on the trays, meaning that $Q_1$ to $Q_N$ in Fig. 11.9 are zero. This is an *adiabatic* column.

In a *diabatic* column, heat may also be added/withdrawn by means of heat exchangers on every tray (*cf.* Fig. 11.9). We shall see that diabatic columns have better thermodynamic efficiencies than adiabatic ones [141, 161, 162, 163]. The total apparatus becomes more complicated, but there are still indications that diabatic distillation may be economically feasible [150, 164].

We shall study binary distillation and use a standard model for the distillation column [165]. The governing equations are given in Table 11.2 [146]. The table gives the total mole balance, the mole balance for the light component, and the balance equation for the internal energy at each tray. Two major assumptions in the model are:

- The total pressure is constant throughout the column.

- The liquid and the vapor that leave a tray, see the small frame in Fig. 11.9, are in equilibrium with each other at the temperature of the tray, $T_n$.

### 11.5.1   The entropy production

The entropy production of the column has separate contributions from all trays. On tray number $n$ there is transport of heat and mass between the liquid and vapor phases. The model contains no details of these processes, and we find the entropy production due to heat and mass transfer across the phase boundary from the entropy balance equation only

$$\left(\frac{dS_{irr}}{dt}\right)_{\text{col},n} = V_n S_n^{\text{V}} + L_n S_n^{\text{L}} - V_{n+1} S_{n+1}^{\text{V}} - L_{n-1} S_{n-1}^{\text{L}} - \frac{Q_n}{T_n} \quad (11.38)$$

The first four terms on the right hand side represent entropy carried in and out with the streams entering and leaving tray $n$. The last term is due to the heat exchanged in the heat exchanger.

There are also contributions to the entropy production from the heat exchangers. We use an average heat transfer force, $X_n$, as described in Table 11.2. The contribution from the heat exchanger on tray $n$ is therefore [143, 166].

$$\left(\frac{dS_{irr}}{dt}\right)_{\text{hx},n} = Q_n X_n \quad (11.39)$$

The entropy production on tray $n$ is the sum of Eqs. (11.38) and (11.39). We want to minimize the total entropy production in the column, which is found by summing over all trays, including the condenser $(n = 0)$ and the reboiler $(n = N + 1)$.

In the next Section, we shall show that the state of minimum entropy production in distillation agrees with the hypothesis for state of minimum entropy production, which was proposed based on reactor results (see Section 11.4.3). In order to explain this, we shall divide the distillation column into five parts; the condenser, the reboiler, the tray(s) where the feed is mixed with the internal streams[1],

---

[1]The tray(s) on which the feed is mixed with the internal streams, depends on whether the feed is a liquid $(n = N_{\text{F}})$, a vapor $(n = N_{\text{F}} - 1)$ or a mixture of both $(n = N_{\text{F}}$ and $N_{\text{F}} - 1)$. See the equations in Table 11.2.

Table 11.2:  Governing equations for stationary state binary tray distillation.

---

**Total mole balance:**

$$V_{n+1} - L_n = \begin{cases} D, & n \in [0, N_F - 2] \\ D - (1 - q) F, & n = N_F - 1 \\ D - F, & n \in [N_F, N + 1] \end{cases}$$

**Mole balance for light component:**

$$V_{n+1}\, y_{n+1} - L_n\, x_n = \begin{cases} D\, x_D, & n \in [0, N_F - 2] \\ D\, x_D - (1 - q)\, F z_F, & n = N_F - 1, \\ D\, x_D - F\, z_F, & n \in [N_F, N + 1]. \end{cases}$$

**Balance equation for internal energy:**

$$Q_n = V_n\, H_n^V + L_n\, H_n^L - V_{n+1}\, H_{n+1}^V - L_{n+1}\, H_{n+1}^L - \kappa$$

$$\text{where } \kappa = \begin{cases} (1 - q)\, F\, H_F^V, & n = N_F - 1, \\ q\, F\, H_F^L, & n = N_F, \\ 0, & \text{otherwise.} \end{cases}$$

**Average force for heat exchange:**

$$X_n = \left(\delta / \lambda_n\, T_n^2\right) (Q_n / A_n)$$

**Balance equation for entropy:**

$$\frac{dS_{\mathrm{irr}}}{dt} = B\, S^B + D\, S^D - F\, S^F + \sum_{n=0}^{N+1} \left(-\frac{Q_n}{T_n} + Q_n\, X_n\right)$$

---

the trays in the rectifying section (above the mixing tray(s)), and the trays in the stripping section (below the mixing tray(s)). Each part has a separate contribution to the total entropy production

$$\left(\frac{dS_{irr}}{dt}\right) = \left(\frac{dS_{irr}}{dt}\right)^{\text{Condens.}} + \left(\frac{dS_{irr}}{dt}\right)^{\text{Reboil.}} + \left(\frac{dS_{irr}}{dt}\right)^{\text{Mix.}}$$

$$+ \sum^{N_{\text{Rectifier}}} \left(\frac{dS_{irr}}{dt}\right)_n + \sum^{N_{\text{Stripper}}} \left(\frac{dS_{irr}}{dt}\right)_n$$

$$(11.40)$$

Furthermore, the five parts of the system will, in general, have different duties, or process intensities. For instance, in the numerical example presented in the next section, the stripping section has to do a more demanding separation than the rectifying section.

The stripping section has therefore a larger contribution to the total entropy production than the rectifying section, both for the adiabatic reference column and the optimized diabatic columns.

## 11.5.2 The state of minimum entropy production

We want to know the properties of the diabatic column with minimum entropy production, especially how it relates to the results for reactors and the hypothesis for the state of minimum entropy production in Section 11.4.3. The optimization problem is to find how the entropy production can be minimized by distributing the heating/cooling capacity over the column in a different way than is done in the adiabatic column. More precisely, we want to find the optimal distributions of heat exchanged and heat transfer area on each tray, $Q_n$ and $A_n$ for $n = 0, 1, ..., N + 1$, when the temperatures/compositions of the feed, distillate and bottom streams, and the total heat transfer area, $A_{\text{total}}$, are fixed. We shall assume here that the heating/cooling utility in the heat exchangers can have any temperature.

We shall use the separation of an equimolar mixture of water and ethanol as an example. The optimization problem can only be solved numerically. The technicalities of the numerical solution are discussed elsewhere [139]. We shall compare the optimal diabatic column with the adiabatic reference column, and a *diabatic EoEP column* [147]. The diabatic EoEP column is a diabatic tray column where additional constraints are introduced in the entropy production minimization; the entropy production is forced to be the same on all trays within a section of the column. We don't force the entropy production to be the same on trays in the stripping section, as on trays in the rectifying section. The reason is that the boundary conditions, and the number of trays in the two sections, dictate different process intensities (duties) in the two sections. A higher process intensity in one section should be reflected in a higher optimal value for the constant tray entropy production in that section. The condenser $(n = 0)$, the reboiler $(n = N + 1)$ and the tray(s) where the feed stream is mixed with the internal streams of the column, are not parts of the stripping or rectifying sections. The tray entropy production in these parts are therefore not constrained to one of the values in the two sections. The heat exchanged and the heat transfer area in these parts of the system are optimized together with the values of the tray entropy production in the stripping and rectifying sections.

The hypothesis for the state of minimum entropy production (see Section 11.4.3), now predicts that *the EoEP column is a good approximation to the true optimum when there is sufficient freedom.* There are many ways to express this freedom in a diabatic distillation column. In this example, we shall discuss the effect of varying the total heat transfer area and the number of trays. The system is most free to adjust when the area and the number of trays are large. The freedom is reduced as both values decrease. The hypothesis predicts therefore, that EoEP approximates the optimum best for large heat transfer areas and/or a large number of trays.

The effects of varying the number of trays and the total heat transfer area on the total entropy production of the adiabatic, optimal and EoEP columns are shown in Figs. 11.10 and 11.11, respectively. The figures show, as expected, that the total entropy production increases when the number of trays and/or the total heat transfer area decrease(s). The adiabatic column has the highest entropy production, and the optimal column has the lowest entropy production. The EoEP column has approximately the same entropy production as the optimal column, except for low heat transfer areas. For low heat transfer areas, the entropy production of the EoEP column is significantly higher than the entropy production of the optimal column. The agreement between the optimal and the EoEP column gets better and better as both the number of trays and/or the total heat transfer area increase(s).

The characteristics of the adiabatic, the optimal and the EoEP columns are presented in Figs. 11.12–11.15. In all the figures, tray 0 is the condenser, and tray 21 is the reboiler. Figures 11.12 and 11.13 present the characteristics of column A. This is a column with 20 trays and a heat transfer area of 20 m$^2$. For this column, the total entropy production of the optimal and the EoEP columns are approximately the same (*cf.* Fig. 11.11).

The vapor flows and the heat duties on each tray in column A are shown in Figs. 11.12 and 11.13, respectively. The characteristics of the adiabatic and the optimal columns are discussed elsewhere

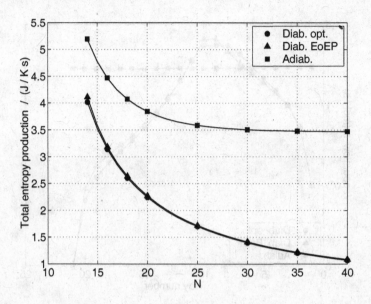

Figure 11.10: The total entropy production vs. the number of trays when the total heat transfer area is $20\,\mathrm{m}^2$.

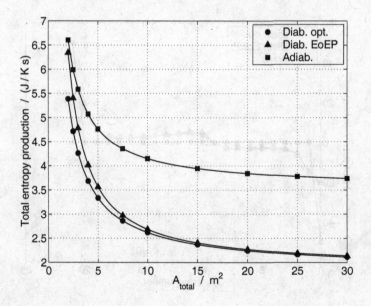

Figure 11.11: The total entropy production vs. the total heat transfer area when there is 20 trays.

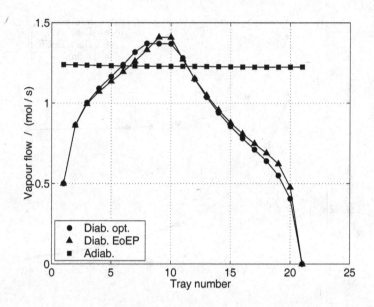

Figure 11.12: The vapor flow on each tray in column A ($N = 20$ trays, heat transfer area is 20 m$^2$).

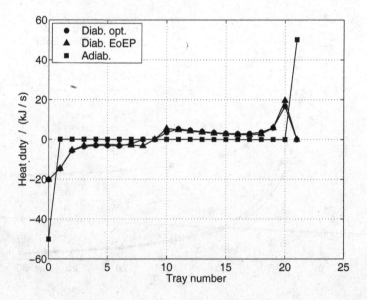

Figure 11.13: The heat duty on each tray in column A ($N = 20$ trays, heat transfer area is 20 m$^2$).

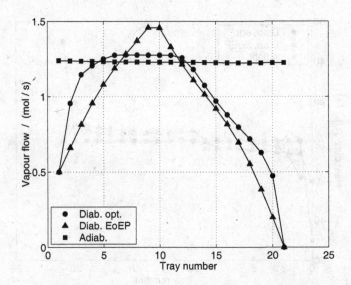

Figure 11.14: The vapor flow on each tray in column B ($N = 20$ trays, heat transfer area is 2 m$^2$).

[143, 139, 144, 145]. The important point here is that the agreement between the optimal column and the EoEP column is very good. The vapor flow on each tray in the EoEP column is shifted slightly down in the rectifying section and slightly up in the stripping section compared to the optimal column, but the differences are not substantial. The differences in the heat duties are barely visible, except on trays 8 and 20.

Figures 11.14 and 11.15 present the characteristics of column B. This is a column with 20 trays and a heat transfer area of 2 m$^2$. For this column, the total entropy production of the optimal and the EoEP columns are significantly different (*cf.* Fig. 11.11).

The profiles of the optimal column B and the EoEP column B are thus also significantly different. This is shown in Figs. 11.14 and 11.15. Fig. 11.14 shows that the vapor flow is too low to be optimal in most of the EoEP column, except close to the feed tray ($n = 9$) where it is too high. The heat duty on each tray in the EoEP column is higher than in the optimal column, especially in the rectifying section (*cf.* Fig. 11.15).

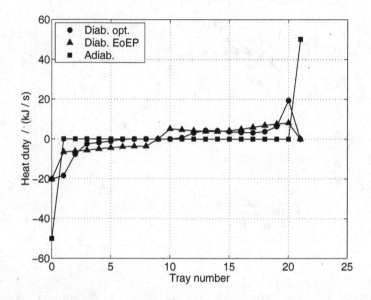

Figure 11.15: The heat duty on each tray in column B ($N = 20$ trays, heat transfer area is $2 \, \mathrm{m}^2$).

The results in Figs. 11.10–11.15 show that the EoEP column is a good approximation to the optimal column, except when the heat transfer area is low (high process intensity). Moreover, the EoEP column approximates the optimal column best for long columns with a large heat transfer area. These results support the hypothesis presented in Section 11.4.3 for the state of minimum entropy production [49]. Using column A and column B as examples, it is clear that there is sufficient freedom to adjust in column A, but not in column B. The fact that the total heat transfer area in column B is one tenth of that in column A, restricts the system's freedom to adjust.

### 11.5.3  Column design

There are two major engineering challenges with making a distillation column fully diabatic. Firstly, the vapor and liquid flows vary throughout the column. This has important implications for the fluid dynamics in the column. To address this, the column with trays can be designed with a varying diameter from top to bottom [167, 168]. The diameter should be roughly proportional with the vapor flow for low-pressure operations. Such a column could be expensive in combination with the costs of extra pumps, extra heat exchangers with tubing, etc. An alternative has been proposed by Olujic and coworkers [169, 170]; the diameter is maintained, while the internal area is varied. In a distillation column which is filled with packing instead of trays, varying vapor and liquid flows can be accommodated with proper choices of packing material. A dense packing with low gas capacity should be used close to the top and the bottom, while a more open packing with higher gas capacity should be used in the middle of the column. In between the packing sections, there are liquid collection and redistribution devices.

The second important challenge is the range of temperatures needed for the utilities. In the optimization, it was assumed that utilities at all temperatures are available in the plant. This is usually not the case in practice. The limited number of available utilities means that one has to approximate the fully diabatic column. One possibility is to use a single heating fluid circulating in series from one tray to the next below the feed tray, and a single cooling fluid circulating in series above the feed tray [171]. The fluid dynamical problem was already mentioned.

One observation for the diabatic column is that the most important heat exchangers are close to the two ends of the column, especially when the column is long [146]. A practical alternative is therefore to only put heat exchangers on trays 1 and N. The hydrodynamics will not be affected too much and the additional capital costs are probably reasonable. Further improvements can be achieved by integrating the two heat exchangers so that the heat added on tray N is taken from tray 1. This can be done with a heat pump [172], or by elevating the pressure in the rectifying section (above the feed tray) in order to shift the phase equilibrium to higher temperatures there [173]. The latter alternative is used in the state-of-the-art technology for air separation [174].

In summary, there are many ways to approach a completely diabatic column. All the options have common characteristics. An energy efficient distillation column allows for heat exchange along the column, facilitated by a distribution of the available heat transfer area. The heat may be exchanged through means of heating/cooling media, or by direct interaction with other columns matching the required heating/cooling duty.

## 11.6    Concluding remarks

In this Chapter, we first explained concepts like work, ideal work, and lost work, and their connection to the entropy production. We have also explained when different optimization problems, like maximum work output and minimum entropy production, are equivalent.

Optimal control theory was introduced in order to solve the entropy production minimization problems using a local description of the entropy production. This robust framework was used to reveal properties of the state of minimum entropy production. Specifically this theory gives optimal time-trajectories of batch processes and optimal process-conditions for stationary state processes. Whereas parameter-optimizations are core activities of process engineers, functional optimizations are relatively new for practical applications and have at the time of writing yet to be explored in many practical areas.

We derived the theorem of equipartition of entropy production for the simple processes; isothermal expansion and heat exchange.

Next, we have studied models of real process units. We have seen that the equipartition results from the simple cases of isothermal expansion and heat exchange do not apply to the entire reactor or distillation column. But, there are parts of these systems which have approximately constant local entropy production and/or thermodynamic forces when the system is in the state of minimum entropy production. The hypothesis for the state of minimum entropy production in an optimally controlled system, explains that we can expect such behavior in (a part of) a system if there is sufficient freedom. Tight constraints lead to deviations from the equipartition principle, however.

We presented the highway in state space for chemical reactors with minimum entropy production. This is a path in the reactor's state space which is crowded by solutions of the optimization problem; in the same way a real highway is crowded by cars. A highway also exists for distillation columns.

From the optimization results, we concluded on a set of guidelines or an energy-efficient operation of process units. Procedures should be further developed to design whole plants, not only process units. It has been proposed that process control can benefit from non-equilibrium thermodynamics [40]. Process control can be used to keep systems on their path of minimum entropy production.

# Appendix A

## A.1 Balance equations for mass, charge, momentum and energy

Balance equations are used in the Gibbs equation in this book, *cf.* Chapters 3 and 6, and (A.39) below, to find the entropy production in terms of the conjugate fluxes and forces. It is the purpose of this Section to give the balance equations in a form that can be used in the Gibbs equation. Balance equations have the general form

$$\text{Accumulation} = \text{Influx} - \text{Outflux} + \text{Production} \tag{A.1}$$

Compare this with Eqs. (3.1) and (3.2)! Prominent balance equations are conservation laws of mass, total energy and momentum. These laws have a zero production term. When the balance equation is formulated for non-conserved properties, like component mass, internal energy or entropy, Eq. (A.1) has a non-zero production term, which may be positive or negative, except for the entropy production which is always positive or zero. Momentum changes are best described when material densities are given in units of $kg/m^3$, while chemical and electrochemical reactions are best described when we used units of $mol/m^3$. Conversions between the two units are therefore often required in the balance equations.

Balance equations for mass and internal energy can be used directly in the Gibbs equation. The momentum balance affects the total energy, however, and enters Gibbs'equation via the internal energy balance. We start with a formulation of the mass balance of integral form, and continue to find the corresponding differential form. We consider only homogeneous systems without relativistic or radiative effects. We do not consider complex fluids, in which there are elastic contributions to the pressure tensor. We shall deal with systems away from mechanical equilibrium and include the effect of electric fields needed for descriptions of electrochemical systems in flow fields, even if this is not elaborated on in this book. Kjelstrup and Bedeaux described heterogeneous systems [31] in mechanical equilibrium.

### A.1.1 Mass balance

The total mass balance is composed of component balances. The balance equation for the mass density $\rho_i$ of component $i$ in a time-independent volume, $V$, is[1]

$$\frac{d}{dt} \int_V \rho_i \, dV = - \int_A \rho_i \mathbf{v}_i \cdot \mathbf{n}_A \, dA + \int_V \nu_i M_i r \, dV \qquad (A.2)$$

where $r$ is the reaction rate in units mol/(m$^3$s), $M_i$ the molecular mass in kg/mol, and $\nu_i$ the stoichiometric coefficient. Furthermore $\mathbf{v}_i$ is the velocity of component $i$. The left-hand side of Eq. (A.2) represents accumulation of mass of component $i$ in volume $V$. The first term on the right-hand side accounts for the net flux of the component through the surface $A$. The vector $\mathbf{n}_A$ in Eq. (A.2) is the normal on the surface. It has a unit length, is orthogonal to the surface, and is pointed outward. The mass-transfer across the surface is obtained using the interior product (or dot-product) of the velocity vector with the normal. The second term on the right-hand side accounts for the net production of the component in a reaction. We restrict ourselves to one reaction. When there are multiple independent reactions, we sum over these reactions. Applying Gauss' divergence theorem, $\int_A \mathbf{F} \cdot \mathbf{n}_A \, dA = \int_V \nabla \cdot \mathbf{F} \, dV$, to the surface integral in Eq. (A.2), using that this equation is valid for an arbitrary choice of the volume, we find the balance equation for $i$ in differential form

$$\frac{\partial \rho_i}{\partial t} = -\nabla \cdot \rho_i \mathbf{v}_i + \nu_i M_i r \qquad (A.3)$$

This equation is often referred to as the continuity equation. Upon summing Eq. (A.3) over all $n$ components, we obtain

$$\frac{\partial \rho}{\partial t} = -\nabla \cdot (\rho \, \mathbf{v}) \qquad (A.4)$$

---

[1]Throughout this appendix we indicate vectors by boldface letters, e.g. the velocity $\mathbf{v}$. The internal product $\mathbf{a} \cdot \mathbf{b} = \sum a_\alpha b_\alpha$, where $\alpha$ is $x, y, z$, leads to a scalar. Analogously, $\nabla \cdot \mathbf{a}$ is the divergence of the vector field $\mathbf{a}$ (and is thus a scalar).

where the total mass density of the mixture is given by

$$\rho \equiv \sum_{i=1}^{n} \rho_i \tag{A.5}$$

The barycentric (center of mass) velocity $\mathbf{v}$ is

$$\mathbf{v} \equiv \frac{1}{\rho} \sum_{i=1}^{n} \rho_i \mathbf{v}_i = \sum_{i=1}^{n} w_i \mathbf{v}_i \tag{A.6}$$

Here $w_i \equiv \rho_i/\rho$ denotes the mass fraction of component $i$. To obtain Eq. (A.4), we used that the total mass is conserved in the reaction

$$\sum_{i=1}^{n} \nu_i M_i = 0 \tag{A.7}$$

Equation (A.4) is the conservation law for the total mass. For the component density $c_i \equiv \rho_i/M_i$ (in $mol/m^3$) the balance equation equivalent to Eq. (A.3) is

$$\frac{\partial c_i}{\partial t} = -\nabla \cdot (c_i \mathbf{v}_i) + \nu_i r \tag{A.8}$$

Summing Eq. (A.8) over all components, we obtain

$$\frac{\partial c}{\partial t} = -\nabla \cdot (c\, \mathbf{v}_{mol}) + r \sum_{i=1}^{n} \nu_i \tag{A.9}$$

for the total molar density of the mixture

$$c \equiv \sum_{i=1}^{n} c_i \tag{A.10}$$

The average molar velocity is defined as

$$\mathbf{v}_{mol} \equiv \frac{1}{c} \sum_{i=1}^{n} c_i \mathbf{v}_i = \sum_{i=1}^{n} x_i \mathbf{v}_i \tag{A.11}$$

Here $x_i \equiv c_i/c$ is the mole fraction of the component. Only Eq. (A.4) for the total mass density is a conservation law. None of the other quantities is conserved. They all satisfy a balance equation with a production term. The last term $r \sum \nu_i$ in Eq. (A.9), for example, is the production of moles in the reaction. Equation (A.9) shall be used in Eq. (A.39) below.

## A.1.2   Momentum balance

The momentum balance is[2]

$$\frac{\partial \rho \mathbf{v}}{\partial t} = -\nabla \cdot (\rho \mathbf{v}\mathbf{v} + \mathbf{P}) + \sum_{i=1}^{n} \rho_i \mathbf{f}_i \qquad (A.12)$$

This equation is also called the equation of motion. The first term on the right hand side is minus the divergence of the convective momentum flux. Further acceleration of the medium is due either to the pressure tensor $\mathbf{P}$, or to external forces $\mathbf{f}_i$ (in units N/kg) acting on component $i$. The pressure tensor can, for isotropic fluids, be decomposed into the static pressure $p$ and a shear contribution $\mathbf{\Pi}$, as

$$\mathbf{P} = p\mathbf{1} + \mathbf{\Pi} \qquad (A.13)$$

where $\mathbf{1}$ is the unit matrix. The momentum balance then becomes

$$\frac{\partial \rho \mathbf{v}}{\partial t} = -\nabla \cdot (\rho \mathbf{v}\mathbf{v} + \mathbf{\Pi}) - \nabla p + \sum_{i=1}^{n} \rho_i \mathbf{f}_i \qquad (A.14)$$

We do not consider complex fluids, in which there are elastic contributions to the pressure tensor. The most common forces $\mathbf{f}_i$ to be considered in Eq. (A.14) are forces due a gravitational field or an electric field. The gravitational field is a conservative field, which is given by

$$\mathbf{f}_i = \mathbf{g} = g(0,0,1) = -\nabla \psi_{grav} = -\nabla gz \qquad (A.15)$$

We assumed the gravitational acceleration $\mathbf{g}$ to be constant and directed in the $z$-direction. By substituting Eq. (A.15) into Eq. (A.14), the equation of motion becomes

$$\frac{\partial \rho \mathbf{v}}{\partial t} = -\nabla \cdot (\rho \mathbf{v}\mathbf{v} + \mathbf{\Pi}) - \nabla p + \rho \mathbf{g} = -\nabla \cdot (\rho \mathbf{v}\mathbf{v} + \mathbf{\Pi}) - \nabla p - \rho \nabla \psi_{grav}$$
$$(A.16)$$

The electric force on component $i$ in the absence of a magnetic field is

$$\mathbf{f}_i = q_i \mathbf{E} = -q_i \nabla \psi \qquad (A.17)$$

---

[2]Throughout this appendix we indicate tensors by boldface capital letters, e.g. for the pressure tensor $\mathbf{P}$. The product $\mathbf{ab}$ of two vectors, with vector-components $a_\alpha$ and $b_\beta$, leads to a tensor with tensor-components $a_\alpha b_\beta$. $\nabla \mathbf{a}$ is the gradient of the vector (and is thus a tensor of rank 2). The internal product of two tensors $\mathbf{A} \cdot\cdot \mathbf{B} = \sum \sum A_{\alpha\beta} B_{\beta\alpha}$ is indicated by a double-dot and gives a scalar.

where $q_i$ is the charge of ion $i$ per unit of mass and $\psi$ is the Maxwell potential. The total charge per unit of mass, $q$, is

$$\rho q = \sum_{i=1}^{n} \rho_i q_i \tag{A.18}$$

By substituting Eqs. (A.17) and (A.18) into Eq. (A.14), the equation of motion becomes

$$\frac{\partial \rho \mathbf{v}}{\partial t} = -\nabla\cdot(\rho\mathbf{v}\mathbf{v} + \mathbf{\Pi}) - \nabla p + \rho z \mathbf{E} = -\nabla\cdot(\rho\mathbf{v}\mathbf{v} + \mathbf{\Pi}) - \nabla p - \rho q \nabla \psi \tag{A.19}$$

There is no magnetic contribution to this equation when the magnetic field is zero. A more detailed discussion was given by de Groot and Mazur [12], Chapter VIII. Most systems relax to an electroneutral state very quickly, in a few nanoseconds. The equation of motion does then, no longer contain an electric force ($q = 0$), and the last term of Eq. (A.19) disappears. We are concerned with such systems in this book. It is then possible to describe the local thermodynamic state by densities of neutral components. For an electrolyte solution, for instance, salt concentrations are sufficient for a complete description. Concentrations of ions can all be given in terms of the salt concentrations. The gradient of the electric potential, $\nabla\phi$, drives the electric flux via the electrodes of the system (the entropy production in Chapter 3 contains the term $-(\mathbf{j}\cdot\nabla\phi)/T$). An electrolyte has an electric current when there is a relative motion of charged components (see e.g. Kjelstrup and Bedeaux [31] Chapters 10 and 17).

$$\frac{\mathbf{j}}{F} = \sum_i z_i \mathbf{J}_i \tag{A.20}$$

where the sum is over the ion fluxes. The charge number $z_i$ is the positive valence number for cations and the negative valence number for anions. The ion flux $J_i$ have contributions to from the electric current and from the appropriate salt flux $J_j$

$$\mathbf{J}_i = \frac{t_i}{z_i}\frac{\mathbf{j}}{F} + \Sigma_j \nu_{ij}\mathbf{J}_j \tag{A.21}$$

where the sum is over the salts and $\nu_{ij}$ is the number of ions of type $i$ in salt $j$. The factor $f_i$ depends on the electrode materials, see example below. We note that $\Sigma_i f_i = 1$ and $\Sigma_i \nu_{ij} z_i = 0$ (salts are electroneutral). When ions are chosen as components, we introduce the Maxwell potential. The electric potential is related to the Maxwell potential $\psi$ (see e.g. Kjelstrup and Bedeaux [31] Chapters 10 and 17) by

$$\phi \equiv \psi + \frac{1}{F} \sum_i \frac{f_i}{z_i} M_i \mu_i = \frac{1}{F} \sum_i \frac{f_i}{z_i} M_i \widetilde{\mu}_i \qquad (A.22)$$

where the sum is over the ions. The electrochemical potential is defined by

$$\widetilde{\mu}_i \equiv \mu_i + \frac{z_i F}{M_i} \psi \qquad (A.23)$$

The chemical and electrochemical potentials are given in J/kg, and $M_i$ is the molar mass in kg/mol. The last identity in Eq. (A.22) gives $\phi$ in terms of electrochemical potentials. In a system with two identical electrodes, the ion involved in the electrode reaction has $f_i = 1$, while the other ions has $f_i = 0$. In a system with two different electrodes and a monovalent salt, $f_i = 1/2$ for each ion involved in the electrode reaction. With two chloride reversible electrodes, say, the electric potential is

$$\phi = \psi - \frac{1}{F} \mu_{Cl^-} = -\frac{1}{F} \widetilde{\mu}_{Cl^-}$$

In a formation cell, with a chloride and a sodium reversible electrode, the relation between $\phi$ and $\psi$ is

$$\phi = \psi + \frac{1}{2F} \mu_{Na^+} - \frac{1}{2F} \mu_{Cl^-} = \frac{1}{2F} \widetilde{\mu}_{Na^+} - \frac{1}{2F} \widetilde{\mu}_{Cl^-}$$

Expressions like these can be used to replace $\phi$ in the energy balance (A.37), if a description in terms of ions is of interest.

### A.1.3   Total energy balance

The total energy of a system is conserved. It can only change if energy is added or removed across the system boundary

$$\frac{\partial \rho e}{\partial t} = -\nabla \cdot \mathbf{J}_e \qquad (A.24)$$

where $e$ is the specific total energy of a system in J/kg, and $\mathbf{J}_e$ is the energy flux in J/(m$^2$s). The total energy per unit of volume $\rho e$ has contributions from internal energy $\rho u$, kinetic energy $\frac{1}{2}\rho v^2$ (where $v = |\mathbf{v}|$), gravitational energy $\rho\psi_{grav}$, and the energy density of the electric field $\frac{1}{2}\varepsilon_0 E^2$ (where $E = |\mathbf{E}|$)

$$\rho e = \rho u + \tfrac{1}{2}\rho v^2 + \rho\psi_{grav} + \tfrac{1}{2}\varepsilon_0 E^2 \tag{A.25}$$

Here $\varepsilon_0$ is the dielectric constant of vacuum. The energy flux can be decomposed as

$$\mathbf{J}_e = \rho\left(u + \tfrac{1}{2}v^2 + \psi_{grav}\right)\mathbf{v} + \mathbf{P}\cdot\mathbf{v} + \frac{\mathbf{j}}{F}\sum_{i=1}^{n}\frac{f_i}{z_i}M_i\mu_i + \mathbf{J}_q \tag{A.26}$$

The first term on the right-hand side of Eq. (A.26) represents the convective energy flux, the second term is the contribution due to mechanical work done on the system, the third term represents the energy flux due to the relative motion of the charged particles. The last term defines the total heat flux.

## A.1.4   Kinetic energy balance

The starting point for deriving a balance equation for the kinetic energy, $\rho v^2/2 = \rho\mathbf{v}\cdot\mathbf{v}/2$, is the momentum balance, Eq. (A.16). Using also the continuity equation, Eq. (A.4), we obtain

$$\frac{\partial}{\partial t}\frac{1}{2}\rho v^2 = \frac{\partial}{\partial t}\frac{1}{2}\rho\mathbf{v}\cdot\mathbf{v} = -\nabla\cdot\left(\tfrac{1}{2}\rho v^2\mathbf{v}\right) - \mathbf{v}\cdot(\nabla\cdot\mathbf{P}) + \rho\mathbf{v}\cdot\mathbf{g} \tag{A.27}$$

The second term on the right hand side of Eq. (A.27) can be rewritten as

$$\mathbf{v}\cdot(\nabla\cdot\mathbf{P}) = \nabla\cdot(\mathbf{P}\cdot\mathbf{v}) - \mathbf{P}:\nabla\mathbf{v} \tag{A.28}$$

By introducing Eq. (A.28) into Eq. (A.27), we obtain a balance equation for the specific kinetic energy

$$\frac{\partial}{\partial t}\frac{1}{2}\rho v^2 = -\nabla\cdot\left(\tfrac{1}{2}\rho v^2\mathbf{v}\right) - \nabla\cdot(\mathbf{P}\cdot\mathbf{v}) + \mathbf{P}:\nabla\mathbf{v} + \rho\mathbf{v}\cdot\mathbf{g} \tag{A.29}$$

## A.1.5   Potential energy balance

The time rate of change of the potential energy density in Eq. (A.25) satisfies

$$\frac{\partial}{\partial t}\rho\psi_{grav} = \psi_{grav}\frac{\partial\rho}{\partial t} = -\psi_{grav}\nabla\cdot\rho\mathbf{v} = -\nabla\cdot\rho\psi_{grav}\mathbf{v} - \rho\mathbf{v}\cdot\mathbf{g} \tag{A.30}$$

where we used that the potential $\psi_{grav}$ does not change with time, and Eqs. (A.4) and (A.15).

### A.1.6   Balance of the electric field energy

To obtain a balance equation for the electric field energy, we need the third Maxwell equation for a zero magnetic field and a zero electric polarization

$$\varepsilon_0 \frac{\partial \mathbf{E}}{\partial t} = -\mathbf{j} \tag{A.31}$$

The balance equation then becomes

$$\frac{\partial}{\partial t} \tfrac{1}{2} \varepsilon_0 E^2 = \varepsilon_0 \mathbf{E} \cdot \frac{\partial \mathbf{E}}{\partial t} = -\mathbf{E} \cdot \mathbf{j} \tag{A.32}$$

### A.1.7   Internal energy balance

The balance equation for the internal energy, which we need for finding the entropy production from Eq. (A.39), is obtained by subtracting Eqs. (A.29), (A.30) and (A.32) from Eq. (A.24) and using Eq. (A.26). We obtain

$$\frac{\partial \rho u}{\partial t} = -\nabla \cdot \left( \rho u \mathbf{v} + \frac{\mathbf{j}}{F} \sum_{i=1}^{n} \frac{f_i}{z_i} M_i \mu_i + \mathbf{J_q} \right) - \mathbf{P} : \nabla \mathbf{v} + \mathbf{E} \cdot \mathbf{j} \tag{A.33}$$

Because the system is electroneutral, $\nabla \cdot \mathbf{j} = 0$. Equation (A.33) becomes

$$\frac{\partial \rho u}{\partial t} = -\nabla \cdot (\rho u \mathbf{v} + \mathbf{J}_q) - \mathbf{P} : \nabla \mathbf{v}$$

$$+ \left( \mathbf{E} - \frac{1}{F} \nabla \sum_{i=1}^{n} \frac{f_i}{z_i} M_i \mu_i \right) \cdot \mathbf{j}$$

$$= -\nabla \cdot (\rho u \mathbf{v} + \mathbf{J}_q) - \mathbf{P} : \nabla \mathbf{v}$$

$$- \left[ \nabla \left( \psi + \frac{1}{F} \sum_{i=1}^{n} \frac{f_i}{z_i} M_i \mu_i \right) \right] \cdot \mathbf{j}$$

$$= -\nabla \cdot (\rho u \mathbf{v} + \mathbf{J}_q) - \mathbf{P} : \nabla \mathbf{v} - (\nabla \phi) \cdot \mathbf{j} \tag{A.34}$$

In many applications of the balance equation for internal energy, it is convenient to decompose the pressure tensor in the static pressure and the shear pressure tensor, according to Eq. (A.13). Considering only the term in Eq. (A.34) containing the pressure tensor

$$-\mathbf{P} : \nabla \mathbf{v} = -p \, \nabla \cdot \mathbf{v} - \mathbf{\Pi} \cdot \nabla \mathbf{v}$$
$$= -\nabla \cdot (p\mathbf{v}) + \mathbf{v} \cdot \nabla p - \mathbf{\Pi} : \nabla \mathbf{v} \qquad (A.35)$$

By introducing Eq. (A.35) into Eq. (A.34), we obtain

$$\frac{\partial \rho u}{\partial t} = -\nabla \cdot \left( \rho(u + \frac{p}{\rho})\mathbf{v} + \mathbf{J}_q \right) + \mathbf{v} \cdot \nabla p - \mathbf{\Pi} : \nabla \mathbf{v} - (\nabla \phi) \cdot \mathbf{j} \quad (A.36)$$

which gives the balance equation of internal energy written in terms of the enthalpy that is carried along, $\rho h = \rho u + p$, as

$$\frac{\partial \rho u}{\partial t} = -\nabla \cdot (\rho h \mathbf{v} + \mathbf{J}_q) + \mathbf{v} \cdot \nabla p - \mathbf{\Pi} : \nabla \mathbf{v} - (\nabla \phi) \cdot \mathbf{j} \qquad (A.37)$$

The internal energy is not conserved, because the last three terms of Eq. (A.37) are source terms. The first is due to mechanical work, the second is due to viscous dissipation and the third is an electric work term. The electric work has its source outside the local volume element. Equation (A.37) is the first law for the systems under consideration. Throughout the text the measurable heat flux $\mathbf{J}_q'$ has been used with $\mathbf{J}_q + \rho h \mathbf{v} = \mathbf{J}_q' + \sum H_i \mathbf{J}_i$.

## A.1.8 Entropy balance

The entropy is not conserved, so the entropy balance is

$$\frac{\partial \rho s}{\partial t} = -\nabla \cdot (\rho s \mathbf{v} + \mathbf{J}_s) + \sigma \qquad (A.38)$$

where $\mathbf{J}_s$ is the entropy flux relative to the barycentric frame of reference and $\sigma$ is the entropy production. According to the second law, $\sigma$ is non-negative. In order to derive explicit expressions for the entropy flux and production, we use the Gibbs relation

$$\frac{\mathrm{d}s}{\mathrm{d}t} = \frac{1}{T} \frac{\mathrm{d}u}{\mathrm{d}t} + \frac{p}{T} \frac{\mathrm{d}v}{\mathrm{d}t} - \sum_{i=1}^{n} \frac{\mu_i}{T} \frac{\mathrm{d}w_i}{\mathrm{d}t} \qquad (A.39)$$

where the specific volume is $v \equiv 1/\rho$ and where the comoving time derivative is defined by

$$\frac{\mathrm{d}}{\mathrm{d}t} \equiv \frac{\partial}{\partial t} + \mathbf{v} \cdot \nabla \qquad (A.40)$$

By introducing the continuity equation (A.4) for the total mass density, we can write for any specific density $a$

$$\rho\frac{da}{dt} = \frac{\partial\rho a}{\partial t} + \nabla\cdot\rho a\mathbf{v} \tag{A.41}$$

Equation (A.38) together with Eq. (A.41) then gives

$$\rho\frac{ds}{dt} = -\nabla\cdot\mathbf{J}_s + \sigma \tag{A.42}$$

From Eq. (A.34) we similarly obtain

$$\rho\frac{du}{dt} = -\nabla\cdot\mathbf{J}_q - \mathbf{P}:\nabla\mathbf{v} - (\nabla\phi)\cdot\mathbf{j} \tag{A.43}$$

and using the continuity equation (A.4) we obtain

$$\rho\frac{dv}{dt} = \nabla\cdot\mathbf{v} \tag{A.44}$$

For the mass fractions we obtain, using Eqs. (A.41) and (A.3),

$$\rho\frac{dw_i}{dt} = \frac{\partial\rho_i}{\partial t} + \nabla\cdot\rho_i\mathbf{v}$$
$$= -\nabla\cdot\rho_i(\mathbf{v}_i - \mathbf{v}) + \nu_i M_i r \tag{A.45}$$

Finally, by substituting Eqs. (A.43)–(A.45) into the Gibbs relation (A.39) we obtain

$$\rho\frac{ds}{dt} = -\frac{1}{T}\nabla\cdot\mathbf{J}_q - \frac{1}{T}\mathbf{\Pi}:\nabla\mathbf{v} - \frac{1}{T}(\nabla\phi)\cdot\mathbf{j}$$
$$+ \sum_{i=1}^{n}\frac{\mu_i}{T}\nabla\cdot\rho_i(\mathbf{v}_i - \mathbf{v}) - \frac{1}{T}\left(\sum_{i=1}^{n}\mu_i\nu_i M_i\right)r \tag{A.46}$$

We define diffusion fluxes relative to the barycentric motion by

$$\mathbf{J}_{i,\text{bar}} \equiv \rho_i(\mathbf{v}_i - \mathbf{v}) \tag{A.47}$$

and the Gibbs energy of the reaction (in J/mol) by

$$\Delta_r G \equiv \sum_{i=1}^{n}\mu_i\nu_i M_i \tag{A.48}$$

Equation (A.46) simplifies to

$$\rho\frac{\mathrm{d}s}{\mathrm{d}t} = -\frac{1}{T}\nabla\mathbf{J}_q - \frac{1}{T}\mathbf{\Pi}:\nabla\mathbf{v} - \mathbf{j}\cdot\frac{\nabla\phi}{T} + \sum_{i=1}^{n}\frac{\mu_i}{T}\nabla\mathbf{J}_{i,\mathrm{bar}} - \frac{\Delta_r G}{T}r \quad (A.49)$$

We rewrite this as

$$\rho\frac{\mathrm{d}s}{\mathrm{d}t} = -\nabla\cdot\frac{\mathbf{J}_q - \sum_{i=1}^{n}\mu_i\mathbf{J}_{i,\mathrm{bar}}}{T} + \mathbf{J}_q\cdot\nabla\frac{1}{T} - \frac{1}{T}\mathbf{\Pi}:\nabla\mathbf{v} - \mathbf{j}\cdot\frac{\nabla\phi}{T}$$
$$-\sum_{i=1}^{n}\mathbf{J}_{i,\mathrm{bar}}\cdot\nabla\frac{\mu_i}{T} - r\frac{\Delta_r G}{T} \quad (A.50)$$

By comparing this equation with Eq. (A.42) we identify the entropy flux

$$\mathbf{J}_s = \frac{\mathbf{J}_q - \sum_{i=1}^{n}\mu_i\mathbf{J}_{i,\mathrm{bar}}}{T} = \frac{\mathbf{J}'_q}{T} + \sum_{i=1}^{n}S_i\mathbf{J}_{i,\mathrm{bar}} \quad (A.51)$$

where $\mathbf{J}'_q$ is the measurable heat flux. The comparison of (A.49) with (A.50) gives the entropy production

$$\sigma = \mathbf{J}_q\cdot\nabla\frac{1}{T} - \frac{1}{T}\mathbf{\Pi}:\nabla\mathbf{v} - \mathbf{j}\cdot\frac{\nabla\phi}{T} - \sum_{i=1}^{n}\mathbf{J}_{i,\mathrm{bar}}\cdot\nabla\frac{\mu_i}{T} - r\frac{\Delta_r G}{T} \quad (A.52)$$

It follows from the definition of the diffusion fluxes that they are not independent. Their sum is zero

$$\sum_{i=1}^{n}\mathbf{J}_{i,\mathrm{bar}} = 0 \quad (A.53)$$

We can therefore preferably write the entropy production in the form

$$\sigma = \mathbf{J}_q\nabla\frac{1}{T} - \frac{1}{T}\mathbf{\Pi}:\nabla\mathbf{v} - \mathbf{j}\cdot\frac{\nabla\phi}{T} - \sum_{i=1}^{n-1}\mathbf{J}_{i,\mathrm{bar}}\nabla\frac{\mu_i - \mu_n}{T} - r\frac{\Delta_r G}{T} \quad (A.54)$$

## A.2 Partial molar thermodynamic properties

Definitions for partial molar properties derive from the Gibbs equation

$$dU = TdS - pdV + \sum_{j=1}^{n} \mu_j dN_j \qquad (A.55)$$

and its integrated form

$$U = TS - pV + \sum_{j=1}^{n} \mu_j N_j \qquad (A.56)$$

The Gibbs-Duhem equation given in Eq. (3.16) is obtained by differentiating the last equation, using the first, and dividing by the volume. The enthalpy is defined as

$$H \equiv U + pV = TS + \sum_{j=1}^{n} \mu_j N_j \qquad (A.57)$$

By using the Gibbs equation, this gives

$$dH = TdS + Vdp + \sum_{j=1}^{n} \mu_j dN_j \qquad (A.58)$$

The Gibbs energy is defined by

$$G \equiv U - TS + pV = \sum_{j=1}^{n} \mu_j N_j \qquad (A.59)$$

Again, by using the Gibbs equation, we obtain

$$dG \equiv -SdT + Vdp + \sum_{j=1}^{n} \mu_j dN_j \qquad (A.60)$$

The partial molar volume and the partial molar entropy are defined by

$$V_j \equiv \left( \frac{\partial V}{\partial N_j} \right)_{p,T,N_k}, \quad S_j \equiv \left( \frac{\partial S}{\partial N_j} \right)_{p,T,N_k} \qquad (A.61)$$

The partial molar properties obey

$$\sum_j N_j V_j = V, \quad \sum_j N_j S_j = S \qquad (A.62)$$

By dividing these relations by the volume, we obtain

$$\sum_j c_j V_j = 1, \qquad \sum_j c_j S_j = \frac{S}{V} = s \qquad \text{(A.63)}$$

From the expression for $dG$ we find alternative expressions for the partial molar quantities

$$V_j \equiv \left( \frac{\partial \mu_j}{\partial p} \right)_{T,N_k}, \qquad S_j \equiv - \left( \frac{\partial \mu_j}{\partial T} \right)_{p,N_k} \qquad \text{(A.64)}$$

This results in the expression for a differential change in the chemical potential

$$
\begin{aligned}
d\mu_j &= -S_j dT + V_j dp + \sum_{k=1}^{n} \left( \frac{\partial \mu_j}{\partial N_k} \right)_{T,p,N_l,E_{eq}} dN_k \\
&= -S_j dT + V_j dp + \sum_{k=1}^{n} \left( \frac{\partial \mu_j}{\partial c_k} \right)_{T,p,c_l,E_{eq}} dc_k \\
&\equiv -S_j dT + V_j dp + d\mu_j^c
\end{aligned}
\qquad \text{(A.65)}
$$

The last line defines $d\mu_j^c$. A combination of terms frequently used in this book, for instance in Eq. (3.15), is

$$d\mu_{j,T} = d\mu_j + S_j dT = V_j dp + d\mu_j^c \qquad \text{(A.66)}$$

where we used the partial molar entropy. From the partial molar quantities defined above we can also define the partial molar internal energy and enthalpy

$$
\begin{aligned}
U_j &= TS_j - pV_j + \mu_j \\
H_j &= TS_j + \mu_j
\end{aligned}
\qquad \text{(A.67)}
$$

These properties are, according to the construction, functions of $p, T$ and $c_k$. The partial molar Gibbs energy is the chemical potential, $G_j = \mu_j$.

**Exercise A.2.1** *Find a form of Gibbs-Duhem's equation, Eq. (3.16), that contains $d\mu_{i,T}$ instead of $d\mu_i$. Assume that the polarization of the system is negligible.*

- **Solution**: The solution is found by substituting Eq. (A.66) into Eq. (3.16):

$$dp = sdT + \sum_{i=1}^{n} c_i d\mu_{i,T} - \sum_{i=1}^{n} c_i S_i dT$$

By using $\sum_{i=1}^{n} c_i S_i = s$, Gibbs-Duhem's equation reduces to

$$dp = \sum_{i=1}^{n} c_i d\mu_{i,T} \qquad (A.68)$$

The equation also applies to a system in a temperature gradient.

## A.3   The chemical potential and its reference states

The chemical potential is a function of temperature, pressure and composition. Like other energy variables, it is not absolute. Only differences in chemical potentials are absolute and can be measured.

The mass transport equations presented, say in Chapter 5, require expressions for the gradient of the chemical potential. For an ideal gas we can calculate this gradient from gradients in temperature, pressure and concentrations, knowing in addition only the ideal gas molar enthalpy $H_i^{ig}(T)$, or heat capacity $c_{p,i}^{ig}(T)$. For real fluids and solids, experimental values or models are needed. There are three main approaches to obtain the chemical potential. We can use

- the equation of state

- the excess Gibbs energy, or

- Henry's law

The first approach is general and suited for any phase, while the last two approaches use additional *auxiliary reference points*: the pure liquid[3] and a pure solvent, respectively. They are for liquid phases, only.

---

[3]For electrolytes, a one-molar solution is commonly chosen as an auxiliary reference point.

All three approaches can be formulated in terms of the ideal gas as a reference. The choice of the considered temperature and pressure of the reference chemical potential is irrelevant when gradients of the chemical potential are targeted and when phase equilibria are calculated. The reference chemical potential drops out of these calculation.

For chemical reactions however, the reference chemical potentials are important. A *standard state* ($p^\ominus = 1$ bar) has been defined in order to allow tabulations of reference state values, usually at 298.15 K.

We summarize here how the three approaches are used to calculate the chemical potential of a single component or a mixture. Calculations for chemical reactions are summarized in Chapter A.4. Superscripts *ig* and *sat* are used as abbreviations for the ideal gas and the saturated pure component (i.e. at vapor-liquid equilibrium), respectively, while subscripts '$V$' and '$L$' are used to denote vapor and liquid phase. Subscript '$0i$' shall denote pure component $i$. A composition dependence is indicated with the symbol $\mathbf{x}$, with $\mathbf{x} = x_1, \ldots x_{n-1}$.

## A.3.1 The equation of state as a basis

The chemical potential of a component $i$ in a real mixture is given by

$$\mu_i(T, p, \mathbf{x}) = \mu_{0i}^{ig}(T, p^*) + RT \ln\left(\frac{f_i}{p^*}\right) \quad \text{(A.69)}$$

This equation defines the fugacity of component $i$ in the given state, $f_i(T, p, \mathbf{x})$. For a pure ideal gas, the fugacity is equal to the pressure. In a mixture of ideal gases, the fugacity of component $i$ is equal to the partial pressure $x_i p$. The pressure $p^*$ can be chosen to be the standard state pressure, $p^* = p^\ominus$, or any pressure, where heat capacity data are available. The choice is irrelevant for phase equilibrium calculations or for chemical potential gradients.

From the fugacity we define the fugacity coefficient, $\phi_i$, with $f_i = x_i \phi_i p$, so that

$$\mu_i(T, p, \mathbf{x}) = \mu_{0i}^{ig}(T, p^*) + RT \ln\left(\frac{x_i \phi_i p}{p^*}\right) \quad \text{(A.70)}$$

The fugacity coefficient $\phi_i(T, p, \mathbf{x})$ accounts for the deviation of the chemical potential from ideal gas behavior, with $\phi_i^{ig} = 1$ for ideal gases. It is always greater than zero. The fugacity coefficient is calculated with an equation of state [175]. Equations of state are nowadays also used for complex fluids, like polymers, associating substances and electrolyte solutions [77, 176, 177, 178, 179, 180]. The equation of state approach can be applied to (coexisting) solid, liquid, and gaseous phases. It is most general, because auxiliary reference points, such as the pure component at vapor pressure or a substance infinitely dilute in a solvent, are not needed.

### A.3.2 The excess Gibbs energy as a basis

The chemical potential of a component in a liquid mixture takes the form

$$\mu_i(T, p, \mathbf{x}) = \mu_i^\circ(T, p, \mathbf{x}^\circ) + RT \ln (x_i \gamma_i) \qquad (A.71)$$

where the choice of the reference chemical potential $\mu_i^\circ$ defines the activity coefficient $\gamma_i$. In most cases, the reference chemical potential is chosen as the pure component chemical potential of the (hypothetical) liquid, $\mu_i^\circ(T, p, \mathbf{x}^\circ) = \mu_{0i}^L(T, p)$ and the activity coefficient accounts for the non-ideality of component $i$ in the mixture at $(T, p, \mathbf{x})$. The reference chemical potential is then conveniently based on the saturated liquid state, as

$$\mu_{0i}^L(T, p) = \mu_{0i}^L(T, p_{0i}^{sat}(T)) + \int_{p_{0i}^{sat}}^{p} V_{0i}^L \, \mathrm{d}p \qquad (A.72)$$

The integral of $V_{0i}^L/RT$ with $p$, is the Poynting correction to the auxiliary reference state (pure liquid at vapor-liquid equilibrium). If the pure substance $i$ is a vapor at $(T, p)$, then the liquid molar volume $V_{0i}^L$ is extrapolated down to $p$. That is practiced in the majority of all phase equilibrium calculations, because the light-boiler's vapor pressure is usually higher than the system pressure.

Often, it is necessary to consider the pure component in the ideal gas state as a reference; for example, to make the approach compatible with Eq. (A.70) for phase equilibrium calculations, or because temperature variations have to be formulated using the ideal gas heat capacity (rather than the liquid state heat capacities). We obtain from Eqs. (A.70) and (A.72) that

$$\mu_{0i}^L(T, p) = \mu_{0i}^{ig}(T, p^*) + RT \ln \left( \frac{\phi_{0i}^{sat} p_{0i}^{sat}}{p^*} \right) + \int_{p_{0i}^{sat}}^{p} V_{0i}^L \, \mathrm{d}p \qquad (A.73)$$

With the ideal gas reference state, the chemical potential is

$$\mu_i(T,p) = \mu_{0i}^{ig}(T,p^*) + RT \ln \left( \frac{x_i \gamma_i \phi_{0i}^{sat} p_{0i}^{sat} \Pi_{0i}}{p^*} \right) \qquad (A.74)$$

The Poynting correction has been estimated here with a constant pure component molar volume, giving

$$\Pi_{0i} = \exp \left( \frac{V_{0i}^L}{RT} \left( p - p_{0i}^{sat} \right) \right) \qquad (A.75)$$

The Poynting correction is near unity at moderate pressure-differences. The fugacity coefficient of the pure saturated vapor phase $\phi_{0i}^{sat}$ is for low pressures approximately unity. The activity coefficients are obtained from excess Gibbs energy models, $(\partial G^E / \partial N_i)_{T,p,N_{j \neq i}} = RT \ln \gamma_i$. In practice, these models are expressed only in terms of temperature and composition as variables, $G^E(T, \mathbf{x})$, and provide the activity coefficients $\gamma_i(T, \mathbf{x})$. Neglecting the pressure leads to numerically simple models. This approach for the chemical potential requires a correlation of the vapor pressure $p_{0i}^{sat}(T)$ and the pure substance $i$ needs to be below the critical point.

### A.3.3 Henry's law as a basis

The solubility of solute $i$ in a solvent $s$ is described by Henry's law in the dilute limit

$$y_i \phi_i^V p = x_i k_{H,i} \qquad (A.76)$$

where Henry's law's constant is defined by $k_{H,i} \equiv \lim_{x_s \to 1} \phi_i^L p$. At finite solute concentration, Henry's law is corrected by an activity coefficient

$$y_i \phi_i^V p = x_i \hat{\gamma}_i \, k_{H,i} \, \Pi_i^\infty \qquad (A.77)$$

The Henry constant in this form is only a function of temperature. The activity coefficient $\hat{\gamma}_i$ is related to the activity coefficient of Eq. (A.74), by $\hat{\gamma}_i = \gamma_i / \gamma_i^\infty$, with $\gamma_i^\infty \equiv \lim_{x_s \to 1} \gamma_i$. The factor $\Pi_i^\infty = \exp \left( \frac{V_i^\infty}{RT} \left( p - p_{0s}^{sat} \right) \right)$ accounts for a deviation in the pressure from the solvent's vapor pressure. Here, the partial molar volume of solute $i$ at infinite dilution, $V_i^\infty$, is used. By introducing Eq. (A.77) into (A.70), we obtain the chemical potential with respect to the ideal gas reference as

$$\mu_i = \mu_{0i}^{ig} + RT \ln \left( \frac{x_i \hat{\gamma}_i k_{H,i} \Pi_i^\infty}{p^*} \right) \qquad (A.78)$$

## A.4   Chemical driving forces and equilibrium constants

For chemical reactions, the reference chemical potentials define the equilibrium constant. It is convention to define the equilibrium 'constant' $K(T)$ as a function of temperature, but at the standard pressure $p^\ominus = 1$ bar. Most tabulated values for the pure component chemical potential (also termed Gibbs energy of formation) are given for the temperature of $T_{298} = 298.15$ K, $\mu_i^\ominus = \mu_{0i}(T_{298}, p^\ominus)$. These values are for a pure component either in *ideal gas state*, or in pure *liquid*, or pure *solid* states. Consistency in the reference state of all components involved is required. As we have seen above, the chemical potential can always be cast into the form

$$\mu_i = \mu_i^{\{\pi\}\ominus} + RT \ln\left(\frac{f_i}{f_i^{\{\pi\}\ominus}}\right) \tag{A.79}$$

where $\pi$ is an index specifying the state, $\pi \in \{ig, L, S\}$ for the ideal gas, liquid or solid state. For the three states, the reference chemical potentials and the corresponding fugacities are

$$\mu_i^{\{ig\}\ominus} = \mu_{0i}^{ig}(T, p^\ominus)$$
$$f_i^{\{ig\}\ominus} = p^\ominus \tag{A.80}$$

$$\mu_i^{\{L\}\ominus} = \mu_{0i}^{L}(T, p^\ominus)$$
$$f_i^{\{L\}\ominus} = \phi_{0i}^{sat} p_{0i}^{sat} \cdot \exp\left(\frac{V_{0i}^{L}}{RT}\left(p^\ominus - p_{0i}^{sat}\right)\right) \tag{A.81}$$

$$\mu_i^{\{S\}\ominus} = \mu_{0i}^{S}(T, p^\ominus)$$
$$f_i^{\{S\}\ominus} = \phi_{0i}^{subl} p_{0i}^{subl} \cdot \exp\left(\frac{V_{0i}^{S}}{RT}\left(p^\ominus - p_{0i}^{subl}\right)\right) \tag{A.82}$$

The driving force for chemical reactions is then, in general,

$$\frac{\Delta_r G}{RT} = \sum_{i=1}^{n} \nu_i \frac{\mu_i}{RT}$$
$$= \underbrace{\sum_{i=1}^{n} \nu_i \frac{\mu_i^{\{\pi\}\ominus}}{RT}}_{\equiv -\ln K^{\{\pi\}}} + \ln\left(\prod_{i=1}^{n}\left(\frac{f_i}{f_i^{\{\pi\}\ominus}}\right)^{\nu_i}\right) \tag{A.83}$$

The equilibrium constant depends thus on the choice of the reference chemical potentials. We give more details below for the case where the equilibrium constant is formulated in terms of ideal gas chemical potentials, and for the case where the equilibrium constant is written in terms of the liquid chemical potential.

## A.4.1   The ideal gas reference state

The driving force for a chemical reaction becomes

$$
\frac{\Delta_r G}{RT} = \underbrace{\sum_{i=1}^{n} \nu_i \frac{\mu_i^{\{ig\}\ominus}}{RT}}_{\equiv -\ln K^{\{ig\}}} + \ln \left( \prod_{i=1}^{n} \left( x_i \phi_i \frac{p}{p^\ominus} \right)^{\nu_i} \right)
\tag{A.84}
$$

This formulation is valid for reactions in any phase. The fugacity coefficient accounts for the deviations from ideal gas state. An equation of state can be used to calculate the fugacity coefficients. The equation shows that driving forces and equilibria of gas-phase reactions are strongly dependent on pressure. The equilibrium constant at any system temperature other than $T_{298} = 298.15\ K$ can be calculated from the integrated Gibbs-Helmholtz relation

$$
\underbrace{\sum_{i=1}^{n} \nu_i \frac{\mu_i^{\{ig\}\ominus}(T)}{RT}}_{\equiv -\ln K^{\{ig\}}} = \underbrace{\sum_{i=1}^{n} \nu_i \frac{\mu_i^{\{ig\}\ominus}(T_{298})}{RT_{298}}}_{\equiv -\ln K_{298}^{\{ig\}}}
$$

$$
- R^{-1} \int_{298.15\ K}^{T} \sum_{i=1}^{n} \nu_i H_i^{ig}(T) \frac{1}{T^2}\ dT
\tag{A.85}
$$

Equations (A.84) and (A.85) require that all the reference chemical potentials are for the ideal gas state. If the chemical potential of any constituent is tabulated for the liquid phase, it can be converted by

$$
\mu_i^{\ominus\{ig\}}(T) = \mu_i^{\ominus\{L\}}(T)
$$

$$
- RT \ln \left( \frac{1}{p^\ominus} p_{0i}^{sat} \phi_{0i}^{sat} \cdot \exp \left( \frac{V_{0i}^L}{RT} \left( p^\ominus - p_{0i}^{sat} \right) \right) \right)
\tag{A.86}
$$

For low enough vapor pressure, $p_{0i}^{sat}$, are the fugacity coefficient and the exponent approximately unity. The argument in the last term in

Eq. (A.86) then simplifies to the vapor-pressure divided by standard pressure only.

## A.4.2    The pure liquid reference state

The equilibrium constant, formulated in terms of liquid phase reference potentials, gives

$$\underbrace{\frac{\Delta_r G}{RT} = \sum_{i=1}^{n} \nu_i \frac{\mu_i^{\{L\}\circ}}{RT} + \ln \left( \prod_{i=1}^{n} \left( x_i \gamma_i \Pi_{0i}^{\circ} \right)^{\nu_i} \right)}_{\equiv -\ln K^{\{L\}}} \tag{A.87}$$

The Poynting correction is modified, *cf.* the regular definition Eq. (A.75), with

$$\Pi_{0i}^{\circ} = \exp \left( \frac{V_{0i}^{L}}{RT} \left( p^{\circ} - p_{0i}^{sat} \right) \right) \tag{A.88}$$

The temperature integration is now carried out in the liquid phase, from tabulated values of the chemical potential at $T_{298}$, so that

$$\underbrace{\sum_{i=1}^{n} \nu_i \frac{\mu_i^{\{L\}\circ}(T)}{RT}}_{\equiv -\ln K^{\{L\}}} = \underbrace{\sum_{i=1}^{n} \nu_i \frac{\mu_i^{\{L\}\circ}(T_{298})}{RT_{298}}}_{\equiv -\ln K_{298}^{\{L\}}}$$

$$- R^{-1} \int_{298.15\,K}^{T} \underbrace{\sum_{i=1}^{n} \nu_i H_{0i}^{L}(T, p^{\circ})}_{\equiv \Delta_r H} \frac{1}{T^2} \, \mathrm{d}T \tag{A.89}$$

where $H_{0i}^{L}$ is the molar enthalpy of the pure liquid. Whenever reference chemical potentials are tabulated for the liquid, then usually the enthalpies are also listed for the liquid. For any component where the enthalpy is tabulated for the ideal gas state, we can calculate the chemical potential difference of Eq. (A.89) using the ideal gas enthalpy. The calculation is done along the integration path: (1) integrate to the ideal gas state at $T_{298} = 298.15\,K$, then (2) integrate from $T_{298}$ to system temperature, and (3) integrate back to the liquid state, so that

$$\frac{\mu_i^{\{L\}\diamond}(T)}{RT} = \frac{\mu_i^{\{L\}\diamond}(T_{298})}{RT_{298}} - R^{-1} \int_{298.15\,K}^{T} H_i^{ig}(T)\frac{1}{T^2}\,\mathrm{d}T$$

$$+ \ln\left(\frac{p_{0i}^{sat}(T)\phi_{0i}^{sat}(T)\exp\left(\frac{V_{0i}^L}{RT}(p^\diamond - p_{0i}^{sat}(T))\right)}{p_{0i}^{sat}(T_{298})\phi_{0i}^{sat}(T_{298})\exp\left(\frac{V_{0i}^L}{RT_{298}}(p^\diamond - p_{0i}^{sat}(T_{298}))\right)}\right)$$

$$(A.90)$$

In many cases (i.e. at low enough vapor pressures), the exponents and the fugacity coefficients can be approximated as unity. The argument in the last term in Eq. (A.90) then simplifies to the vapor-pressure ratio.

## A.5    Minimizing the total entropy production of a $K$-step expansion process

Isothermal expansion is discussed in Chapters 11.1 and 11.2. The $K$-step expansion process with minimum entropy production is illustrated in Fig. 11.2 (b)–(d) for $K{=}3$, 5 and 15 steps, respectively. Here, we solve this optimization problem analytically.

The optimization problem is to minimize the total entropy production of the $K$-step expansion process Eq. (11.8). In each step, the time variation of the gas pressure is given by Eq. (11.1). The initial and final gas pressures of the overall process are $p_1$ and $p_2$, respectively.

By integrating Eq. (11.1) over step $i$, we obtain

$$p_{\text{ext},i} = \frac{1}{\alpha - 1}\left(\alpha\,p_{1,i} - p_{1,i+1}\right), \qquad i \in [1, K] \qquad (A.91)$$

where $\alpha = \exp\left(-\frac{f}{N\,R\,T_0}\frac{\Theta}{K}\right)$. We used here that $p_{\text{ext},i}$ is constant, but different from $p_{2,i} = p_{1,i+1}$ in each step. By introducing this result into Eq. (11.8), we obtain

$$\frac{dS_{\text{irr}}}{dt} = -N\,R\,\ln\left(\frac{p_2}{p_1}\right) - \frac{N\,R}{\alpha - 1}\sum_{i=1}^{K}\left(\alpha\,\frac{p_{1,i}}{p_{1,i+1}} - \alpha - 1 + \frac{p_{1,i+1}}{p_{1,i}}\right)$$

$$(A.92)$$

We minimize the total entropy production subject to $p_{1,1} = p_1$ and $p_{1,K+1} = p_2$, and construct therefore the Euler-Lagrange function

$$\mathcal{L} = -N\,R\,\ln\left(\frac{P_2}{P_1}\right) - \frac{N\,R}{\alpha - 1} \sum_{i=1}^{K}\left(\alpha\,\frac{p_{1,i}}{p_{1,i+1}} - \alpha - 1 + \frac{p_{1,i+1}}{p_{1,i}}\right)$$
$$+ \lambda_1\,(p_{1,1} - p_1) + \lambda_2\,(p_{1,K+1} - p_2) \tag{A.93}$$

from which we derive the necessary conditions for the optimum:

$$0 = \frac{\partial\mathcal{L}}{\partial p_{1,i}} = \begin{cases} -\frac{N\,R}{\alpha - 1}\left(\alpha\,\frac{1}{p_{1,2}} - \frac{p_{1,2}}{p_{1,1}^2}\right) + \lambda_1,\, i = 1 \\[2mm] -\frac{N\,R}{\alpha - 1}\left(\alpha\,\frac{1}{p_{1,i+1}} - \frac{p_{1,i+1}}{p_{1,i}^2} - \alpha\,\frac{p_{1,i-1}}{p_{1,i}^2} + \frac{1}{p_{1,i-1}}\right),\, i \in [2, K] \\[2mm] -\frac{N\,R}{\alpha - 1}\left(-\alpha\,\frac{p_{1,K}}{p_{1,K+1}^2} + \frac{1}{p_{1,K}}\right) + \lambda_2,\, i = K + 1 \end{cases}$$
$$\tag{A.94}$$

By solving these conditions, we find the optimal initial pressure of the gas and the optimal external pressure in each step:

$$p_{1,i} = p_1\left(\frac{p_2}{p_1}\right)^{(i-1)/K}, \qquad\qquad i \in [1, K+1] \tag{A.95}$$

$$p_{\text{ext},i} = p_1\left(\frac{p_2}{p_1}\right)^{(i-1)/K}\left(\frac{\alpha - (p_2/p_1)^{1/K}}{\alpha - 1}\right), \qquad i \in [1, K] \tag{A.96}$$

The corresponding total entropy production is

$$\frac{dS_{\text{irr}}}{dt} = -N\,R$$
$$\times\left[\ln\left(\frac{p_2}{p_1}\right) + \frac{K}{\alpha - 1}\left(\alpha\left(\frac{p_2}{p_1}\right)^{-\frac{1}{K}} - \alpha - 1 + \left(\frac{p_2}{p_1}\right)^{\frac{1}{K}}\right)\right]$$
$$\tag{A.97}$$

and the corresponding work and lost work are:

$$w = -N\,R\,T_0\,\frac{K}{\alpha - 1}\left(\alpha\left(\frac{p_2}{p_1}\right)^{-\frac{1}{K}} - \alpha - 1 + \left(\frac{p_2}{p_1}\right)^{\frac{1}{K}}\right) \tag{A.98}$$

$$w_{\text{lost}} = -N\,R\,T_0$$
$$\times\left[\ln\left(\frac{p_2}{p_1}\right) + \frac{K}{\alpha - 1}\left(\alpha\left(\frac{p_2}{p_1}\right)^{-\frac{1}{K}} - \alpha - 1 + \left(\frac{p_2}{p_1}\right)^{\frac{1}{K}}\right)\right]$$
$$\tag{A.99}$$

The fomulae are illustrated in Fig. 11.2.

## A.6   The work produced by a heat exchanger

Concepts like work, ideal work and lost work are not as obvious for heat exchange as they are for the expansion in Section 11.1. We shall therefore explain that by cooling a hot fluid stream, the heat exchange process produces work. Likewise, by heating a cold fluid stream, the heat exchange process consumes work. A heat exchanger is thus a work-producing or work-consuming apparatus. By convention, we define work done and heat added as positive when they increase the internal energy of a system.

Figure A.1 shows how work production by heat exchange can be seen conceptually. At each position $z$, the cold fluid, with temperature $T_c(z)$, is connected to a reservoir at $T_0$ through a Carnot machine. In the small element $dz$, the heat which is transferred from the cold fluid to the hot fluid, is $dq = \Delta y \, J_q'(z) \, dz$. This heat is supplied to the cold stream by taking the heat $dq_0 = \frac{T_0}{T_c(z)} \, dq$ from the environment using the Carnot machine. The work needed to be done. This is

$$ dw = \eta_C \left( \Delta y \, J_q'(z) \, dz \right) = \Delta y \left( 1 - \frac{T_0}{T_c(z)} \right) J_q'(z) \, dz \qquad \text{(A.100)} $$

where $\eta_C = 1 - T_0/T_c(z)$ is the Carnot efficiency[4]. The total work requirement of the heat exchange process is therefore

$$ w = \Delta y \int_0^L \left( 1 - \frac{T_0}{T_c(z)} \right) J_q'(z) \, dz $$

$$ = q - \Delta y \, T_0 \int_0^L \frac{J_q'(z)}{T_c(z)} \, dz \qquad \text{(A.101)} $$

$$ = F_{\text{out}} \, H_{\text{out}} - F_{\text{in}} \, H_{\text{in}} - \Delta y \, T_0 \int_0^L \frac{J_q'(z)}{T_c(z)} \, dz $$

In the last equality, we used the first law which gives

$$ F_{\text{out}} \, H_{\text{out}} - F_{\text{in}} \, H_{\text{in}} = q \qquad \text{(A.102)} $$

---

[4]The heat $dq$ is the sum of the work needed in the Carnot machine and the heat extracted from the environment, that is $dq = \frac{T_0}{T_c(z)} \, dq + \eta_C \, dq$.

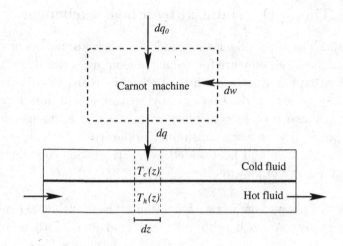

Figure A.1: The heat exchanger as a work-producing or work-consuming machine

where $H_{\text{out}}$ and $H_{\text{in}}$ are the molar enthalpy of the hot fluid out and in, respectively, and $q$ is the total heat transferred to the hot fluid:

$$q = \Delta y \int_0^L J_q'(z)\, dz \qquad (A.103)$$

Equation (A.101) shows that the work requirement is negative when we cool the hot fluid stream. This means that work can in fact be extracted from the process. The opposite is true when we consider heating a cold stream.

**Remark 12** *Compare now Eqs. (11.26) and (A.101). We see again (cf. Section 11.1.3) that minimizing the total entropy production and minimizing the work requirement (maximizing the work output) are equivalent problems when the inlet and outlet states of the hot fluid stream are fixed ($S_{\text{in}}$, $S_{\text{out}}$, $H_{\text{in}}$ and $H_{\text{out}}$ are fixed). Both optimization problems reduce to*

$$\min \qquad \left( -\int_0^L \frac{J_q'(z)}{T_c(z)}\, dz \right) \qquad (A.104)$$

*The outcome of both optimization problems is to take heat out ($J_q'(z) < 0$) at as high temperatures as possible, and to supply heat ($J_q'(z) > 0$)*

*at as low temperatures as possible. In other words, we should use low quality heat for heating and extract heat with high quality, a well-known strategy in process integration. When the inlet and/or the outlet states are not fixed, the optimization problems are not equivalent. We study entropy production minimization because the entropy production is the most fundamental measure of energy dissipation.*

We do not propose that a heat exchanger should be operated as a work-producing or work-consuming machine. In reality, heat extracted from a hot fluid stream is used for other purposes elsewhere in the plant. The same goes for the heat required to heat a cold fluid stream.

## A.7   Equipartition theorems

We shall show that Equipartition of Entropy Production (EoEP) and Equipartition of Forces (EoF) characterize the state of minimum entropy production when some assumptions are fulfilled. The equipartition theorems can be derived both for linear and for non-linear flux-force relations, but slightly different sets of assumptions are needed in the two cases.

The entropy production minimization problems which we discuss in Chapter 11 have a general form. We shall generalize the problem using a matrix formulation. We organize the $N$ state varables in a vector $\mathbf{y}$ and the $M$ control variables in a vector $\mathbf{u}$. The balance equations, the total entropy production and the Hamilton for the optimal control problem can be written as

$$\frac{d\mathbf{y}}{dz} = \mathbf{A}(\mathbf{y}) \, \boldsymbol{\Gamma} \, \mathbf{J}(\mathbf{y}, \mathbf{u}, \mathbf{x}(\mathbf{y}, \mathbf{u})), \qquad (A.105)$$

$$\left(\frac{dS}{dt}\right)_{\text{irr}} = \int_0^L \mathbf{x}(\mathbf{y}, \mathbf{u})^{\mathrm{T}} \, \boldsymbol{\Gamma} \, \mathbf{J}(\mathbf{y}, \mathbf{u}, \mathbf{x}(\mathbf{y}, \mathbf{u})) \, dz, \qquad (A.106)$$

and

$$H(\mathbf{y}, \mathbf{u}, \lambda) = \mathbf{x}(\mathbf{y}, \mathbf{u})^{\mathrm{T}} \boldsymbol{\Gamma} \mathbf{J}(\mathbf{y}, \mathbf{u}, \mathbf{x}(\mathbf{y}, \mathbf{u})) + \lambda^{\mathrm{T}} \mathbf{A}(\mathbf{y}) \boldsymbol{\Gamma} \mathbf{J}(\mathbf{y}, \mathbf{u}, \mathbf{x}(\mathbf{y}, \mathbf{u})), \qquad (A.107)$$

respectively. Here, $\mathbf{A}(\mathbf{y})$ is a matrix with proportionality factors, and $\boldsymbol{\Gamma}$ is a diagonal matrix with geometric factors, $\mathbf{J}(\mathbf{y}, \mathbf{u}, \mathbf{x}(\mathbf{y}, \mathbf{u}))$ is the flux vector, $\mathbf{x}(\mathbf{y}, \mathbf{u})$ is the force vector, and $\boldsymbol{\lambda}^{\mathrm{T}}$ is the vector with multiplier functions. The most general dependence on the state vector $\mathbf{y}$ and the the control vector $\mathbf{u}$ is shown in Eqs. (A.105)–(A.107).

The general formulation describes evolution in one dimension, time or space. We have chosen to use the spatial coordinate $z$ here. This means that the system is stationary. All equations and results in this appendix hold, however, equally well if we switch from space to time as variable. We are thus not restricted to stationary systems.

Many models of engineering systems fit into this form. Among stationary systems, some examples are plug flow reactors, heat exchangers, packed columns used for distillation, absorption or extraction, and membrane processes. The batch counterparts of these systems are examples of time-dependent systems.

We shall go on to derive EoEP and EoF using the general matrix formulation. But first, we give an example of the matrices and the vectors in Eqs. (A.105)–(A.107). For this we use the plug flow reactor problem with one reaction (see Section 11.4). There are $N = 3$ state variables and an $M = 1$ control variable in this problem. We organize these variables in the state vector, $\mathbf{y} = [T, P, \xi_1]^{\mathrm{T}}$, and the control vector, $\mathbf{u} = [T_{\mathrm{a}}]$. We find the matrices $\mathbf{A}(\mathbf{y})$ and $\boldsymbol{\Gamma}$ by comparison of Eqs. (A.105)–(A.107) with Table 11.1:

$$\mathbf{A}(\mathbf{y}) = \begin{bmatrix} \frac{1}{\sum_i F_i C_{p,i}} & 0 & \frac{-\Delta_r H_1}{(\sum_i F_i C_{p,i})} \\ 0 & -\frac{f}{\Omega} & 0 \\ 0 & 0 & \frac{1}{F_A^0} \end{bmatrix}, \quad \boldsymbol{\Gamma} = \begin{bmatrix} \pi D & 0 & 0 \\ 0 & \Omega & 0 \\ 0 & 0 & \Omega \rho_{\mathrm{B}} \end{bmatrix}$$

(A.108)

where $f$ in $\mathbf{A}(\mathbf{y})$ is

$$f = \left( \frac{150\,\mu}{D_{\mathrm{p}}^2} \frac{(1-\epsilon)^2}{\epsilon^3} + \frac{1.75\,\rho^0\,v^0}{D_{\mathrm{p}}} \frac{1-\epsilon}{\epsilon^3} \right)$$

Furthermore, $\mathbf{J}(\mathbf{y}, \mathbf{u}, \mathbf{x}(\mathbf{y}, \mathbf{u}))$ and $\mathbf{x}(\mathbf{y}, \mathbf{u})$ are vectors that contain the fluxes and the forces, respectively:

$$\mathbf{J}(\mathbf{y}, \mathbf{u}, \mathbf{x}(\mathbf{y}, \mathbf{u})) = [J_q, v, r_1]^{\mathrm{T}}$$

(A.109)

and

$$\mathbf{x}\left(\mathbf{y},\mathbf{u}\right) = \left[\left(\frac{1}{T} - \frac{1}{T_{\mathrm{a}}}\right), \left(-\frac{1}{T}\frac{dP}{dz}\right), \left(-\frac{\Delta_{\mathrm{r}}G_1}{T}\right)\right]^{\mathrm{T}} \quad \text{(A.110)}$$

We have indicated in Eqs. (A.105)–(A.110) that $\mathbf{\Gamma}$ is a constant matrix, $\mathbf{A}$ depends on the state vector, and $\mathbf{J}$ and $\mathbf{x}$ depend on both the state vector and the control vector.

In order to derive EoEP and EoF, we use the necessary conditions for minimum entropy production, which can now be written as

$$\frac{d\mathbf{y}}{dz} = \left(\frac{\partial H}{\partial \boldsymbol{\lambda}}\right)^{\mathrm{T}} \qquad \frac{d\boldsymbol{\lambda}}{dz} = -\left(\frac{\partial H}{\partial \mathbf{y}}\right)^{\mathrm{T}} \quad \text{(A.111)}$$

and

$$\frac{\partial H}{\partial \mathbf{u}} = 0 \quad \text{(A.112)}$$

where the derivatives of $H$ ($\frac{\partial H}{\partial \boldsymbol{\lambda}}$, $\frac{\partial H}{\partial \mathbf{y}}$ and $\frac{\partial H}{\partial \mathbf{u}}$) are row vectors. In addition to this, the Hamiltonian is autonomous, as before, and is thus constant along the $z$-coordinate [156].

We assume here that the weak form in Eq. (A.112) is sufficient, see discussion below Eq. (11.14). This is the only assumption we make in order to arrive at Eqs. (A.111) and (A.112) in addition to the assumptions within the model of the system. The necessary conditions in Eqs. (A.111) and (A.112) and the fact that the Hamiltonian is constant are therefore mathematically robust. The details of the state of minimum entropy production are not obvious, though. We shall see below, how physical insight into the solution can be gained by introducing assumptions in a stepwise manner.

## The Spirkl-Ries quantity

We start by making the assumption of enough control variables:

1) There are enough control variables to control all the forces independently and without any constraints on their values. This means that there are at least $N$ control variables ($M \geq N$).

This assumption makes it possible to use **x** as the control instead of **u**. We can then eliminate **u** from the problem and obtain a simpler Hamiltonian

$$H\left(\mathbf{y}, \mathbf{x}, \lambda\right) = \mathbf{x}^{\mathrm{T}} \, \mathbf{\Gamma} \, \mathbf{J}\left(\mathbf{y}, \mathbf{x}\right) + \lambda^{\mathrm{T}} \, \mathbf{A}\left(\mathbf{y}\right) \, \mathbf{\Gamma} \, \mathbf{J}\left(\mathbf{y}, \mathbf{x}\right) \qquad \text{(A.113)}$$

The forces appear now explicitly in $H$, and implicitly in the flux relations. The necessary conditions, equivalent to Eq. (A.112), become:

$$\frac{\partial H}{\partial \mathbf{x}} = \left(\mathbf{\Gamma} \mathbf{J}\left(\mathbf{y}, \mathbf{x}\right)\right)^{\mathrm{T}} + \mathbf{x}^{\mathrm{T}} \frac{\partial \left(\mathbf{\Gamma} \mathbf{J}\left(\mathbf{y}, \mathbf{x}\right)\right)}{\partial \mathbf{x}} + \lambda^{\mathrm{T}} \mathbf{A}(\mathbf{y}) \frac{\partial \left(\mathbf{\Gamma} \mathbf{J}\left(\mathbf{y}, \mathbf{x}\right)\right)}{\partial \mathbf{x}} = 0$$
$$\text{(A.114)}$$

By solving this equation for $\lambda^{\mathrm{T}} \mathbf{A}$ and introducing the result in Eq. (A.113), we obtain

$$H = -\left(\mathbf{\Gamma} \mathbf{J}\left(\mathbf{y}, \mathbf{x}\right)\right)^{\mathrm{T}} \left(\frac{\partial \left(\mathbf{\Gamma} \mathbf{J}\left(\mathbf{y}, \mathbf{x}\right)\right)}{\partial \mathbf{x}}\right)^{-1} \left(\mathbf{\Gamma} \mathbf{J}\left(\mathbf{y}, \mathbf{x}\right)\right) \quad \text{(A.115)}$$

The combination of geometric factors and fluxes, and their derivative, on the right-hand side of Eq. (A.115), is constant along the $z$-coordinate. This was called the Spirkl-Ries quantity by [181] because these authors were the first who proved the result [151]. The Spirkl-Ries quantity is valid for any flux-force relation, given that the number of control variables is at least as high as the number of state variables. It has no simple meaning unless the flux-force relations are linear, It then reduces to EoEP and EoF as discussed below.

## EoEP for linear flux-force relations

To the result above, we can now add the assumption that is standard in irreversible thermodynamics [12, 19]

2) The flux-force relations are linear, that is $\mathbf{J}\left(\mathbf{y}, \mathbf{x}\right) = \mathbf{L}\left(\mathbf{y}\right) \mathbf{x}$.

Here, $\mathbf{L}(\mathbf{y})$ is the matrix with conductivities. With this assumption we obtain $\partial \left(\mathbf{\Gamma} \mathbf{J}\left(\mathbf{y}, \mathbf{x}\right)\right) / \partial \mathbf{x} = \mathbf{\Gamma} \mathbf{L}(\mathbf{y})$ and a reduction of Eq. (A.115) to $H = -\sigma$. In other words: The local entropy production is constant along the $z$-coordinate when assumptions 1) and 2) are valid. This is the theorem of equipartition of entropy production (EoEP) which has been reported by many authors [182, 153, 152, 135, 151, 150].

## EoF for linear flux-force relations

Another result in literature is the theorem of equipartition of forces, EoF [153, 154]. In order to obtain EoF, we first make the assumption of constant conductivities

3) The conductivity matrix, $\mathbf{L}$, does not depend on $\mathbf{y}$ and is therefore constant.

Assumption 3) reduces EoEP to EoF immediately for a system with one force. In the general case $(N > 1)$, the fact that $\sigma = \mathbf{x}^{\mathrm{T}} \boldsymbol{\Gamma} \mathbf{L} \mathbf{x}$ is constant, makes one combination of the forces, but not necessarily each force, constant. In order to derive EoF when $N > 1$, we first rearrange Eq. (A.114) using assumption 2) and obtain

$$\mathbf{x}^{\mathrm{T}} = -\frac{1}{2} \boldsymbol{\lambda}^{\mathrm{T}} \mathbf{A} \qquad (A.116)$$

This equation shows that all forces are constant if $\boldsymbol{\lambda}^{\mathrm{T}} \mathbf{A}$ is constant. It is possible that $\boldsymbol{\lambda}^{\mathrm{T}}$ and $\mathbf{A}$ vary in such a way that the product is constant, but in general both $\boldsymbol{\lambda}^{\mathrm{T}}$ and $\mathbf{A}$ must be constant. In order for $\boldsymbol{\lambda}^{\mathrm{T}}$ to be constant, the Hamiltonian cannot depend on $\mathbf{y}$ (see the right part of Eq. (A.111)). We therefore need to make a fourth assumption in order to prove EoF for the general case $(N > 1)$:

4) The matrix $\mathbf{A}$ does not depend on $\mathbf{y}$ and is therefore constant.

When assumptions 1) to 4) hold, the constant Hamiltonian reduce to EoF. The derivation of EoEP required only assumptions 1) and 2). EoEP is therefore generally a better approximation to the constant Hamiltonian.

## EoEP and EoF for nonlinear flux-force relations

We can also derive EoEP and EoF for nonlinear flux-force relations. More precisely, we consider flux-force relations where the fluxes are functions of the forces only. Thus, we keep assumption 1), and we replace assumptions 2) and 3) with

$2^*$) The fluxes do not depend on the state vector $\mathbf{y}$, that is $\mathbf{J} = \mathbf{J}(\mathbf{x})$.

When this is the case, the local entropy production (*cf.* Eq (A.106)) is only a function of the forces, meaning that the local entropy production is constant (EoEP) whenever all the forces are constant (EoF).

A system with one flux/one force is a special case. In this case, EoF and EoEP follow directly from the Spirkl-Ries quantity, Eq. (A.115), when $\mathbf{J} = \mathbf{J}(\mathbf{x})$. The reason is that the right hand side of Eq. (A.115) is then only a function of the single force in the system. The functional form of $\mathbf{J}(\mathbf{x})$ does not matter, meaning that the flux-force relation can be nonlinear. When there is more than one flux / one force, the situation is more complicated. For $\mathbf{J} = \mathbf{J}(\mathbf{x})$, the Spirkl-Ries quantity says that a function of the forces should be constant. This does not imply, however, that all the forces, and therefore also not the entropy production, are constant.

In order to obtain EoF and EoEP for $N > 1$ fluxes/forces, we must assume that the matrix $\mathbf{A}$ is constant, i.e. assumption 4). This means that the Hamiltonian does not depend on $\mathbf{y}$ any more (*cf.* Eq. (A.107)). The second necessary condition in Eq. (A.111) then gives that $\lambda$ is constant. So, when $\mathbf{A}$ and $\lambda$ are constant, Eq. (A.114) is a set of $N$ algebraic equations in the $N$ forces. Since the forces are independent, the solution of the algebraic equations is that all forces, and the local entropy production, are constant (EoEP and EoF). Once again, the functional form of $\mathbf{J}(\mathbf{x})$ does not matter, and the flux-force relation can be nonlinear.

We have shown that the state of minimum entropy production is characterised by EoEP and EoF when assumption 1), $2^*$) and 4) apply. To this end it should be noted that the combination of assumptions 2) and 3) is one version of assumption $2^*$). In other words, the derivation of EoF for linear flux-force relations is a subset of the more general derivation given here. But, EoEP for linear flux-force relations is more general because only assumptions 1) and 2) were needed.

In entropy production minimization problems of industrial relevance, there is usually not enough control variables to control all forces in the system independently. This means that assumption 1) is not fulfilled, and no equipartition results can be derived, regardless if all the other assumptions are fulfilled. Therefore, it is surprising that the highway in state space for chemical reactors is approximated well by EoEP and EoF (see Section 11.4.3).

# Bibliography

[1] W. Thomson (Lord Kelvin). *Mathematical and Physical Papers. Collected from different Scientific Periodicals from May, 1841, to the Present Time*, volume II. Cambrigde University Press, London, 1884.

[2] L. Onsager. Reciprocal relations in irreversible processes. I. *Phys. Rev.*, 37:405–426, 1931.

[3] L. Onsager. Reciprocal relations in irreversible processes. II. *Phys. Rev.*, 38:2265–2279, 1931.

[4] H. Hemmer, H. Holden and S.K. Ratkje, editor. *The Collected Works of Lars Onsager*. World Scientific, Singapore, 1996.

[5] J. Meixner. Zur thermodynamik der thermodiffusion. *Ann. Physik 5. Folge*, 39:333–356, 1941.

[6] J. Meixner. Reversible bewegungen von flüssigkeiten und gasen. *Ann. Physik 5. Folge*, 41:409–425, 1942.

[7] J. Meixner. Zur Thermodynamik der irreversibelen Prozesse in Gasen mit chemisch reagierenden, dissozierenden und anregbaren Komponenten. *Ann. Physik 5. Folge*, 43:244–270, 1943.

[8] J. Meixner. Zur Thermodynamik der Irreversibelen Prozesse. *Zeitschr. Phys. Chem. B*, 53:235–263, 1943.

[9] I. Prigogine. *Etude thermodynamique des phenomenes irreversibles*. Desoer, Liege, 1947.

[10] P. Mitchell. Coupling of phosphorylation to electron and hydrogen transfer by a chemi-osmotic type of mechanism. *Nature (London)*, 191:144–148, 1961.

[11]  S. R. de Groot and P. Mazur. *Non-Equilibrium Thermodynam-ics.* North-Holland, Amsterdam, 1962.

[12]  S. R. de Groot and P. Mazur. *Non-Equilibrium Thermodynam-ics.* Dover, London, 1984.

[13]  R. Haase. *Thermodynamics of Irreversible Processes.* Addison-Wesley, Reading, MA, 1969.

[14]  R. Haase. *Thermodynamics of Irreversible Processes.* Dover, London, 1990.

[15]  A. Katchalsky and P. Curran. *Nonequilibrium Thermodynam-ics in Biophysics.* Harvard University Press, Cambrigde, Mas-sachusetts, 1975.

[16]  S.R. Caplan and A. Essig. *Bioenergetics and linear nonequilib-rium thermodynamics — The steady state.* Harvard University Press, Cambrigde, Massachusetts, 1983.

[17]  Y. Demirel. *Nonequilibrium thermodynamics.* Elsevier, Boston, 2002.

[18]  K. S. Førland, T. Førland and S. Kjelstrup Ratkje. *Irreversible thermodynamics. Theory and application.* Wiley, Chichester, 1988.

[19]  K. S. Førland, T. Førland and S. Kjelstrup. *Irreversible ther-modynamics. Theory and application.* Tapir, Trondheim, 3rd. edition, 2001.

[20]  V.P. Carey. *Statistical Thermodynamics and Microscale Ther-mophysics.* Cambrigde University Press, Cambridge, 1999.

[21]  D. Kondepudi and I. Prigogine. *Modern thermodyamics. From heat engines to dissipative structures.* Wiley, Chichester, 1998.

[22]  D. Jou, J. Casas-Vasquez and G. Lebon. *Extended Irreversible Thermodynamics.* Springer; Berlin, 2 edition, 1996.

[23]  H.C. Øttinger. *Beyond Equilibrium Thermodynamics.* Wiley-Interscience, Hoboken, 2005.

[24] D.D. Fitts. *Nonequilibrium Thermodynamics*. McGraw-Hill, New York, 1962.

[25] G.D.C. Kuiken. *Thermodynamics for irreversible processes*. Wiley, Chichester, 1994.

[26] A. Perez-Madrid I. Pagonabarraga and J.M. Rubi. Fluctuating hydrodynamics approach to chemical reactions. *Physica A*, 237:205–219, 1997.

[27] I. Pagonabarraga and J.M. Rubi. Derivation of the Langmuir adsorption equation from non-equilibrium thermodynamics. *Physica A*, 188:553–567, 1992.

[28] D. Reguera and J.M. Rubi. Non-equilibrium translational-rotational effects in nucleation. *J. Chem. Phys.*, 115:7100–7106, 2001.

[29] C.M. Guldberg and P. Waage. Studies concerning affinity. *Forhandlinger: Videnskabs-Selskabet i Christiania*, page 35, 1864.

[30] D. Bedeaux and P. Mazur. Mesoscopic non-equilibrium thermodynamics for quantum systems. *Physica A*, 298:81–100, 2001.

[31] S. Kjelstrup and D. Bedeaux. *Non-equilibrium Thermodynamics of Heterogeneous Systems. Series on Advances in Statistical Mechanics. Vol.16*. World Scientific, Singapore, 2008.

[32] K.G. Denbigh. The second-law efficiency of chemical processes. *Chem. Eng. Sci.*, 6:1–9, 1956.

[33] R. B. Bird, W. E. Stewart and E. N. Lightfoot. *Transport Phenomena*. Wiley, 1960.

[34] R. Taylor and R. Krishna. *Multicomponent Mass Transfer*. Wiley, New York, 1993.

[35] E.L. Cussler. *Diffusion, Mass Transfer in Fluid Systems*. Cambridge, 2nd. edition, 1997.

[36] R. Krishna and J.A. Wesselingh. The Maxwell-Stefan approach to mass transfer. *Chem. Eng. Sci.*, 52:861–911, 1997.

[37] A. Bejan. Entropy generation minimization: The new thermodynamics of finite-size devices and finite-time processes. *J. Appl. Phys.*, 79:1191–1218, 1996.

[38] J. Szargut, D.R. Morris and F. R. Steward. *Exergy Analysis of Thermal, Chemical and Metallurgical Processes*. Hemisphere, New York, 1988.

[39] R.S. Berry, V. Kazakov, S. Sieniutycz, Z. Szwast and A.M. Tsirlin. *Thermodynamic Optimization of Finite-Time Processes*. Wiley, Chichester, 2000.

[40] W.L. Luyben, B. Tyreus, and M.L. Luyben. *Plantwide Process Control*. McGraw-Hill, New York, 1998.

[41] D. Bedeaux and S. Kjelstrup. Impedance spectroscopy of surfaces described by irreversible thermodynamics. *J. Non-Equilib. Thermodyn.*, 24:80–96, 1999.

[42] G. Tsatsaronis. Definition and nomenclature in exergy analysis and exergoeconomics. In S. Kjelstrup, J. Hustad, T. Gundersen, A. Røsjorde and G. Tsatsaronis, editor, *Proceedings of ECOS 2005*, pages 321–326, Trondheim, Norway, June 20 - June 22 2005. Norwegian University of Science and Technology, Norway. ISBN 82-519-2041-8.

[43] A. Valero and A. Valero. *Thanatia. The Destiny of the Earths Mineral Resources*. World Scientific, 2015.

[44] J.W. Gibbs. *Collected Works, 2 vols*. Dover, London, 1961.

[45] A. Bejan. *Entropy Generation Minimization. The Method of Thermodynamic Optimization of Finite-Size Systems and Finite-Time Processes*. CRC Press, New York, 1996.

[46] M. J. Moran and H. N. Shapiro. *Fundamentals of Engineering Thermodynamics*. Wiley, New York, 2nd. edition, 1993.

[47] A. Zvolinschi and S. Kjelstrup. A process maturity indicator for industrial ecology. *J. Ind. Ecol.*, 12:159–172, 2008.

[48] P.W. Atkins. *Physical Chemistry*. Oxford, 6th. edition, 1998.

[49] E. Johannessen and S. Kjelstrup. Minimum entropy production in plug flow reactors: An optimal control problem solved for SO$_2$ oxidation. *Energy*, 29:2403–2423, 2004.

[50] A. Zvolinschi, E. Johannessen and S. Kjelstrup. The second-law optimal operation of a paper drying machine. *Chem. Eng. Sci.*, 61:3653–3662, 2006.

[51] B. Hafskjold and S. Kjelstrup Ratkje. Criteria for local equilibrium in a system with transport of heat and mass. *J. Stat. Phys.*, 78:463–494, 1995.

[52] T. Holt, E. Lindeberg and S. Kjelstrup Ratkje. The effect of gravity and temperature gradients on methane distribution in oil reservoirs. *SPE-paper no. 11761*, page 19, 1983.

[53] L.J.T.M. Kempers. A thermodynamic theory of the Soret effect in a multicomponent liquid. *J. Chem. Phys.*, 90:6541–6548, 1989.

[54] A.P. Fröba, S. Will, Y. Nagasaka, J. Winkelmann, S. Wiegand, W. Köhler. *Experimental Thermodynamics Volume IX: Advances in Transport Properties of Fluids*, chapter 2. Optical Methods. Royal Society of Chemistry, 2014.

[55] W.H. Furry, R.C. Jones and L. Onsager. On the theory of isotope separation by thermal diffusion. *Phys. Rev.*, 55:1083–1095, 1939.

[56] C. Debuschewitz and W. Köhler. Molecular Origin of Thermal Diffusion in Benzene + Cyclohexane Mixtures. *Phys. Rev. Letters*, 87:055901–1, 2001.

[57] L.J.T.M. Kempers. A comprehensive thermodynamic theory of the Soret effect in a multicomponent gas, liquid or solid. *J. Chem. Phys.*, 115:6330–6341, 2001.

[58] V.E. Zinoviev. *Thermophysical properties of metals at high temperatures*. Metallurgiya, Moscov, 1989.

[59] H.S. Harned and B.B. Owen. *Physical Chemistry of Electrolytic Solutions*. Reinhold, New York, 3rd. edition, 1958.

[60] S. Kjelstrup and A. Røsjorde. Local and total entropy production and heat and water fluxes in a one-dimensional polymer electrolyte fuel cell. *J. Phys. Chem. B*, 109:9020–9033, 2005.

[61] X. Kang, M. T. Børset, O.S. Burheim, G.M. Haarberg, Q. Xu, S. Kjelstrup. Seebeck coefficients of cells with molten carbonates relevant for the metallurgical industry. *Electrochim. Acta*, 182:342–350, 2015.

[62] X. Liu, S.K. Schnell, J.-M. Simon, P. Krüger, D. Bedeaux, S. Kjelstrup, A. Bardow, and T. J. H. Vlugt. Diffusion coefficients from molecular dynamics simulations in binary and ternary mixtures. *Int. J. Thermodynamics*, 34:1169–1196, 2013.

[63] X. Liu, A. Martin-Calvo, E. Garrity, S.K. Schnell, S. Calero, J.-M. Simon, D. Bedeaux, S. Kjelstrup, A. Bardow, and T. J. H. Vlugt. Fick diffusion coefficients in ternary liquid systems from equilibrium molecular dynamics simulations. *Ind. Eng. Chem. Res.*, 51:10247–10258, 2012.

[64] X. Liu, S.K. Schnell, J.-M. Simon, D. Bedeaux, S. Kjelstrup, A. Bardow, and T.J.H. Vlugt. Fick diffusion coefficients of liquid mixtures directly obtained from equilibrium molecular dynamics. *J. Phys. Chem. B*, 115:12921–12929, 2011.

[65] J. Meixner. Strömungen von fluiden Medien mit inneren Umwandlungen und Druckviscosität. *Zeitschrift für Physik*, 131:456–469, 1952.

[66] S. Hess. Irreversible thermodynamics of nonequilibrium alignment phenomena in molecular liquids and in liquid crystals. I. *Z. Naturforschung*, 30a:728, 1975.

[67] S. Hess. Irreversible thermodynamics of nonequilibrium alignment phenomena in molecular liquids and in liquid crystals. II. viscous flow and flow alignment in the isotropic (stable ad metastable) and nematic phases. *Z. Naturforschung*, 30a:1224, 1975.

[68] D. Bedeaux and J.M. Rubi. Nonequilibrium thermodynamics of colloids. *Physica A*, 305:360–370, 2002.

[69] D. Frenkel and B. Smit. *Understanding molecular simulation: from algorithms to applications*. Academic Press, 2001.

[70] M. Stavrou E. Sauer and J. Gross. Comparison between a homo-and a heterosegmented group contribution approach based on the perturbed-chain polar statistical associating fluid theory equation of state. *Ind. Eng. Chem. Res.*, 53(38):14854–14864, 2014.

[71] O. Lötgering-Lin and J. Gross. Group contribution method for viscosities based on entropy scaling using the perturbed-chain polar statistical associating fluid theory. *Ind. Eng. Chem. Res.*, 54(32):7942–7952, 2015.

[72] Y. Rosenfeld. A quasi-universal scaling law for atomic transport in simple fluids. *J. Phys.: Condens. Matter*, 11(28):5415–5427, 1999.

[73] Y. Rosenfeld. Relation between the transport coefficients and the internal entropy of simple systems. *Phys. Rev. A*, 15:2545–2549, 1977.

[74] J.O. Hirschfelder, C.F. Curtiss, R.B. Bird and M.G. Mayer. *Molecular theory of gases and liquids*, volume 26. Wiley New York, 1954.

[75] L.T. Novak. Fluid viscosity-residual entropy correlation. *Int. J. Chem. Reactor Eng.*, 9(1), 2011.

[76] G. Galliero and C. Boned. Thermal conductivity of the lennard-jones chain fluid model. *Phys. Rev. E*, 80:061202, 2009.

[77] J. Gross and G. Sadowski. Perturbed-chain saft: An equation of state based on a perturbation theory for chain molecules. *Ind. Eng. Chem. Res.*, 40(4):1244–1260, 2001.

[78] M. Hopp and J. Gross. Thermal conductivity of real substances using PCP-SAFT and excess entropy scaling. *Ind. Eng. Chem. Res.*, page submitted, 2016.

[79] J. Xu, S. Kjelstrup, D. Bedeaux and J.-M. Simon. Transport properties of $2F = F_2$ in a temperature gradient as studied by

molecular dynamics simulations. *Phys. Chem. Chem. Phys.*, 9:969–981, 2007.

[80] H. Eyring and E. Eyring. *Modern Chemical Kinetics*. Chapman & Hall, London, 1965.

[81] J.M. Rubi and S. Kjelstrup. Mesoscopic nonequilibrium thermodynamics gives the same thermodynamic basis to Butler-Volmer and Nernst equations. *J. Phys. Chem. B*, 107:13471–13477, 2003.

[82] S. Kjelstrup, J.M. Rubi and D. Bedeaux. Energy dissipation in slipping biological pumps. *Phys. Chem. Chem. Phys.*, 7:4009–4018, 2005.

[83] H.A. Kramers. Brownian motion in a field of force and the diffusion model of chemical reactions. *Physica*, 7:284–304, 1940.

[84] B. Welch. Aluminum Production Paths in the New Millennium. *J. Metals*, pages 24–28, 1999.

[85] R. Huglen and H. Kvande. Global considerationf of aluminium electrolysis on energy and the environment. *Light Metals*, pages 373–380, 1994.

[86] K. Grjotheim and B. Welch. *Aluminium Smelter Technology*. Aluminium-Verlag, Düsseldorf, 2nd. edition, 1988.

[87] T. S. Sørensen and S. Kjelstrup. Paralell Butler-Volmer reactions at the carbon anode in a laboratory cell for electrolytic reduction of aluminium, one producing CO and another producing $CO_2$. II. Effect of temperature and relations between current efficiency, carbon consumption and gas composition. *Aluminium Trans.*, 1:186–196, 1999.

[88] T. S. Sørensen and S. Kjelstrup. Paralell Butler-Volmer reactions at the carbon anode in a laboratory cell for electrolytic reduction of aluminium, one producing CO and another producing $CO_2$. I. Electrode kinetic model of the carbon anode and its application on current-voltage curves. *Aluminium Trans.*, 1:179–185, 1999.

[89] E. M. Hansen and S. Kjelstrup. Application of nonequilibrium thermodynamics to the electrode surfaces of aluminium electrolysis cells. *J. Electrochem. Soc.*, 143:3440, 1996.

[90] E. M. Hansen. *Modeling of Aluminium Electrolysis Cells Using Non-Equilibrium Thermodynamics.* PhD thesis, University of Leiden, 1997.

[91] J. Hives, J. Thonstad, Å. Sterten and P. Fellner. Electrical conductivity of molten cryolite-based mixtures obtained with a tube-type cell made of pyrolitic boron nitride. *Light Metals*, page 187, 1994.

[92] E. M. Hansen, E. Egner and S. Kjelstrup. Peltier effects in electrode carbon. *Metall. and Mater. Trans. B*, 29:69–76, 1997.

[93] K. Grjotheim and H. Kvande. *Understanding the Hall-Heroult Process for Production of Aluminium.* Aluminium-Verlag, Düsseldorf, 1986.

[94] D. Bedeaux and S. Kjelstrup. Transfer coefficients for evaporation. *Physica A*, 270:413–426, 1999.

[95] D. Bedeaux, L.F.J. Hermans and T. Ytrehus. Slow evaporation and condensation. *Physica A*, 169:263–280, 1990.

[96] D. Bedeaux and S. Kjelstrup. Irreversible Thermodynamics - a Tool to describe Phase Transitions far from Global Equilibrium. *Chem. Eng. Sci.*, 59:109–118, 2004.

[97] D. Bedeaux and S. Kjelstrup. Heat, mass and charge transport and chemical reactions at surfaces. *Int. J. of Thermodynamics*, 8:25–41, 2005.

[98] D. Bedeaux, A.M. Albano and P. Mazur. Boundary conditions and nonequilibrium thermodynamics. *Physica A*, 82:438–462, 1976.

[99] A.M. Albano and D. Bedeaux. Non-equilibrium electro-thermodynamics of polarizable multicomponent fluids with an interface. *Physica*, A147:407–435, 1987.

[100] S. Kjelstrup and D. Bedeaux. *Experimental Thermodynamics Volume X: Non-equilibrium Thermodynamics with Applications*, chapter Electrochemical Energy Conversion, pages 244–270. Royal Society of Chemistry, 2016.

[101] D. Bedeaux and S. Kjelstrup. *Experimental Thermodynamics Volume X: Non-equilibrium Thermodynamics with Applications*, chapter Non-equilibrium Thermodynamics for Evaporation and Condensation, pages 154–177. Royal Society of Chemistry, 2016.

[102] D. Bedeaux. Nonequilibrium thermodynamics and statistical physics of surfaces. *Adv. Chem. Phys.*, 64:47–109, 1986.

[103] J.S. Rowlinson and B. Widom. *Molecular Theory of Capillarity*. Oxford, 1982.

[104] T. Savin, K.S. Glavatskiy, S. Kjelstrup, H.C. Öttinger and D. Bedeaux. Exploring the property of local equilibrium for the Gibbs surface in two-phase multi-component mixtures. *Eur. Phys. Letters*, 97:40002–7, 2012.

[105] A.M. Albano, D. Bedeaux and J. Vlieger. On the description of interfacial properties unsing sigular densities and currents at a dividing surface. *Physica A*, 99:293–304, 1979.

[106] A.M. Albano, D. Bedeaux and J. Vlieger. On the description of interfacial electromagnetic properties using singular fields, charge density and currents at a dividing surface. *Physica A*, 102A:105–119, 1980.

[107] S. Kjelstrup and D. Bedeaux. Jumps in electric potential and in temperature at the electrode surfaces of the solid oxide fuel cell. *Physica A*, 244:213–226, 1997.

[108] Christoph Klink, Christian Waibel, and Joachim Gross. Analysis of interfacial transport resistivities of pure components and mixtures based on density functional theory. *Industrial & Engineering Chemistry Research*, 54(45):11483–11492, 2015.

[109] Ø. Wilhelmsen, T.T. Trinh, S. Kjelstrup, T.S. van Erp and D. Bedeaux. Heat and mass transfer across interfaces in complex nanogeometries. *Phys. Rev. Letters*, 114:065901, 2015.

[110] Ø. Wilhelmsen, T. T. Trinh, A. Lervik, V. K. Badam, S. Kjelstrup and D. Bedeaux. Interface transfer coefficients for condensation and evaporation of water. *Phys. Rev. E*, 93:032801, 2016.

[111] F.E. Genceli Güner, M. Rodriguez Pascual, S. Kjelstrup, and G.-J. Witkamp. Coupled Heat and Mass Transfer during Crystallization of $MgSO_4$ $7H_2O$ on a Cooled Surface. *Crystal Growth & Design*, 9:1318–1326, 2009.

[112] F.E. Güner, J. Wåhlin, M. Hinge, and S. Kjelstrup. The temperature jump at a growing ice-water interface. *Chem. Phys. Letters*, 622:15–19, 2015.

[113] Guggenheim E.A. The conceptions of electrical potential difference between two phases and the individual activities of ions. *J. Phys. Chem.*, 33:842–849, 1928.

[114] Guggenheim E.A. *Thermodynamics*. North Holland, Amsterdam, 1985.

[115] J.S. Newman. *Electrochemical Systems*. Prentice-Hall, Englewood Cliffs, 2nd. edition, 1991.

[116] D. Bedeaux, H.C. Öttinger, and S. Kjelstrup. Nonlinear coupled equations for electrochemical cells as developed by the general equation for nonequilibrium reversible-irreversible coupling. *J. Chem. Phys.*, 141:124102, 2014.

[117] M. Findlay. Vaporization through porous membranes,. *Ind. Eng. Process Design and Development*, 6:226, 1967.

[118] A. Jansen, J. Assink, J. Hanemaaijer, J. van Medervoort, E. van Sonsbeek. Development and pilot testing of full-scale membrane distillation modules for deployment of waste heat. *Desalination*, 323:55–65, 2013.

[119] J.H. Hanemaaijer. A method of converting thermal energy into mechanical energy, and an apparatus therefore. *wO Patent App. PCT/NL2012/000,018*, 2013.

[120] G. Scatchard. Ion exchange electrodes. *J. Am. Chem. Soc.*, 75:2883–2887, 1953.

[121] L. Keulen, L.V. van der Ham, T.J.H. Vlugt, N.J.M. Kuipers, S. Kjelstrup. Membrane distillation against a pressure difference, *J. Membr. Science*, accepted, 10.1016/j.memsci.2016. 10.054.

[122] M. Ottøy. *Mass and Heat Transfer in Ion-Exchange Membranes, dr. ing. Thesis no. 50*. University of Trondheim, Norway, 1996.

[123] J. Fischbarg. Fluid Transport Across Leaky Epithelia: Central Role of the Tight Junction and Supporting Role of Aquaporins. *Physiol. Rev.*, 90:1271–1290, 2010.

[124] J. Benavente and C. Fernandez-Pineda. Electrokinetic phenomena in porous membranes: determination of phenomenological coefficients and transport numbers. *Journal of Membrane Science*, 23:121–136, 1985.

[125] T. Okada, S. Kjelstrup Ratkje, H. Hanche-Olsen . Water transport in cation-exchange membranes. *J. Membr. Sci.*, 66:179–192, 1992.

[126] T. Okada, S. Kjelstrup Ratkje, S. Møller-Holst, L.O. Jerdal, K. Friestad, G. Xie and R. Holmen. Water and ion transport in the cation-exchange membrane systems NaCl-SrCl$_2$ and KCl-SrCl$_2$. *J. Membr. Sci.*, 111:159–167, 1996.

[127] O. Gorseth T. Okada, S. Møller-Holst and S. Kjelstrup. Transport and equilibrium properties of nafion membranes with H$^+$ and Na$^+$-ions. *J. Electroanal. Chem.*, 442:137–145, 1998.

[128] T. Okada, G. Xie, O. Gorseth, S. Kjelstrup, N. Nakamura and T. Arimura. Ion and water transport characteristics of nafion membranes as electrolytes. *Electrochim. Acta*, 43:3741–3747, 1998.

[129] P. Trivijitkasem and T. Østvold. Water transport in ion exchange membranes. *Electrochim. Acta*, 25:171–178, 1980.

[130] M. Ottøy, T. Førland, S. Kjelstrup Ratkje and S. Møller-Holst. Membrane transference numbers from a new emf method. *J. Membr. Sci.*, 74:1–8, 1992.

[131] T.S. Brun and D. Vaula. Correlation of measurements of electroosmosis and streaming potentials in ion exchange membranes. *Ber. Bunsenges. Physik. Chem.*, 71:824–829, 1967.

[132] O.S. Burheim, F. Seland, J.G. Pharoah and S. Kjelstrup. Improved electrode systems for reverse electro-dialysis and electrodialysis. *Desalination*, 285:147–152, 2012.

[133] O.S. Burheim, J. Pharoah, D. Vermaas, B.B. Sales, K. Nijmeijer and H.V.M. Hamelers. *Reverse Electrodialysis as an Electric Power Plant*, volume I, page 1482. Wiley, New York, 2013.

[134] L. Nummedal and S. Kjelstrup. Equipartition of forces as a lower bound on the entropy production in heat transfer. *Int. J. Heat Mass Transfer*, 44:2827–2833, 2000.

[135] E. Johannessen and L. Nummedal and S. Kjelstrup. Minimizing the entropy production in heat exchange. *Int. J. Heat Mass Transfer*, 45:2649–2654, 2002.

[136] E. Johannessen and S. Kjelstrup. A highway in state space for reactors with minimum entropy production. *Chem. Eng. Sci*, 60:3347–3361, 2005.

[137] L. Nummedal and M. Costea and S. Kjelstrup. Minimizing the entropy production rate of an exothermic reactor with constant heat transfer coefficient: The ammonia reaction. *Ind. Eng. Chem. Res.*, 42:1044–1056, 2003.

[138] L. Nummedal, A. Røsjorde, E. Johannessen, and S. Kjelstrup. Second law optimisation of a tubular steam reformer. *Chem. Eng. Process*, 44:429–440, 2005.

[139] A. Røsjorde and E. Johannessen and S. Kjelstrup. Minimising the entropy production rate in two heat exchangers and a reactor. In N. Houbak, B. Elmegaard, B. Qvale, and M.J. Moran, editors, *Proceedings of ECOS 2003*, pages 1297–1304, Copenhagen, Denmark, June 30 - July 2 2003. Department of Mechanical Engineering, Technical University of Denmark. ISBN 87-7475-297-9.

[140] Ø. Wilhelmsen, E. Johannessen and S. Kjelstrup. Energy efficient reactor design simplified by second law analysis. *Int. J. Hydrogen Energy*, 35:13219–13231, 2010.

[141] G. M. de Koeijer and S. Kjelstrup. Minimizing entropy production in binary tray distillation. *Int. J. Appl. Thermodyn.*, 3:105–110, 2000.

[142] G.M. de Koeijer and S. Kjelstrup and P. Salamon and G. Siragusa and M. Schaller and K.H. Hoffmann. Comparison of entropy production rate minimization methods for binary diabatic tray distillation. *Ind. Eng. Chem. Res.*, 41:5826–5834, 2002.

[143] G.M. de Koeijer and A. Røsjorde and S. Kjelstrup. Distribution of heat exchange in optimum diabatic distillation columns. *Energy*, 29:2425–2440, 2004.

[144] M. Schaller and K. H. Hoffmann and G. Siragusa and P. Salamon and B. Andresen. Numerically optimized performance of diabatic distillation columns. *Comp. Chem. Eng.*, 25:1537–1548, 2001.

[145] M. Schaller and K. H. Hoffmann and R. Rivero and B. Andresen and P. Salamon. The influence of heat transfer irreversibilities on the optimal performance of diabatic distillation columns. *J. Non-Equilib. Thermodyn.*, 27:257–269, 2002.

[146] A. Røsjorde and S. Kjelstrup. The second law optimal state of a diabatic binary tray distillation column. *Chem. Eng. Sci.*, 60:1199–1210, 2005.

[147] E. Johannessen and A. Røsjorde. Equipartition of entropy production as an approximation to the state of minimum entropy production in a diabatic distillation column. *Energy*, 32:467–473, 2007.

[148] J. Humphrey and A. Siebert. Separation technologies: An opportunity for energy savings. *Chemical Engineering Progress*, 88:32–41, 1992.

[149] A. Røsjorde, S. Kjelstrup, E. Johannessen and R. Hansen. Minimizing the entropy production in a chemical process for dehydrogenation of propane. *Energy*, 32:335–343, 2007.

[150] D. Tondeur and E. Kvaalen. Equipartition of entropy production. An optimality criterion for transfer and separation processes. *Ind. Eng. Chem. Res.*, 26:50–56, 1987.

[151] W. Spirkl and H. Ries. Optimal finite-time endoreversible processes. *Phys. Rev. E*, 52:3485–3489, 1995.

[152] L. Diosi and K. Kulacsy and B. Lukacs and A. Racz. Thermodynamic length, speed, and optimum path to minimize entropy production. *J. Chem. Phys.*, 105:11220–11225, 1996.

[153] D. Bedeaux, F. Standaert, K. Hemmes and S. Kjelstrup. Optimization of processes by equipartition. *J. Non-Equilib. Thermodyn.*, 24:242–259, 1999.

[154] E. Sauar, S. Kjelstrup and K. M. Lien. Equipartition of forces. A new principle for process design and operation. *Ind. Eng. Chem. Res.*, 35:4147–4153, 1996.

[155] P. W. Atkins. *Physical Chemistry*. Oxford, 5th. edition, 1994.

[156] A.E. Bryson and Y.C. Ho. *Applied Optimal Control. Optimization, estimation and control*. Wiley, New York, 1975.

[157] L.S. Pontryagin and V.G. Boltyanskii and R.V. Gamkrelidze and E.F. Mishchenko. *The Mathematical Theory of Optimal Processes*. Pergamon Press, Oxford, 1964.

[158] Ø. Wilhelmsen, E. Johannessen and S. Kjelstrup. *Experimental Thermodynamics Volume X: Non-equilibrium Thermodynamics with Applications*, chapter Entropy Production Minimization with Optimal Control Theory, pages 271–289. Royal Society of Chemistry, 2016.

[159] I.L. Leites, D.A. Sama and N. Lior. The theory and practice of energy saving in the chemical industry: some methods for reducing thermodynamic irreversibility in chemical technology processes. *Energy*, 28:55–97, 2003.

[160] G.M. de Koeijer and E. Johannessen and S. Kjelstrup. The second law optimal path of a four-bed $SO_2$ converter with five heat exchangers. *Energy*, 29:526–549, 2004.

[161] R. Rivero. *L'analyse d'exergie: Application à la Distillation et aux Pompes à Pompes à Chaleur à Absorption*. PhD thesis, Institut National Polytechnique de Lorraine, Nancy, France, 1993.

[162] P. Le Goff, T. Cachot and R. Rivero. Exergy analysis of distillation processes. *Chem. Eng. Technol.*, 19:478–485, 1996.

[163] R. Agrawal and Z.T. Fidbowski. On the use of intermediate reboilers in the rectifying section and condensers in the stripping section of a distillation column. *Ind. Eng. Chem. Res.*, 35(8):2801–2807, 1996.

[164] S. Kauchali, C. McGregor and D. Hildebrandt. Binary distillation re-visited using the attainable region theory. *Comp. Chem. Eng.*, 24:231–237, 2000.

[165] W. McCabe and J. Smith and P. Harriot. *Unit Operations of Chemical Engineering*. McGraw-Hill, New York, 5th. edition, 1993.

[166] G.M. de Koeijer and R. Rivero. Entropy production and exergy loss in experimental distillation columns. *Chem. Eng. Sci.*, 58:1587–1597, 2003.

[167] Z. Fonyo. Thermodynamic analysis of rectification. I. Reversible model of rectification. *Int. Chem. Eng.*, 14:18–27, 1974.

[168] G.M. De Koeijer. *Energy Efficient Operation of Distillation Columns and a Reactor Applying Irreversible Thermodynamics*. PhD thesis, Norwegian University of Science and Technology, Department of Chemistry, Trondheim, Norway, 2002. ISBN 82-471-5436-6, ISSN 0809-103x.

[169] J. de Graauw, A. de Rijke, Z. Olujic and P.J. Jansens. Distillation column with heat integration. *European Patent no.*, EP1332781:06–08, 2003.

[170] Z. Olujic, L. Sun, A. de Rijke and P.J. Jansens. Conceptual design of energy efficient propylene splitter. In R. Rivero, L. Monroy, R. Pulido and G. Tsatsaronis, editor, *Proc. of ECOS 2004*. Instituto Mexicano del Petroleo, Mexico, 2004. ISBN 968-489-027, pages 61–78.

[171] E.S. Jimenez and P. Salamon and R. Rivero and C. Rendon and K.H. Hoffmann and M. Schaller and B. Andresen. Optimization of a Diabatic Distillation Column with Sequential Heat Exchangers. *Ind. Eng. Chem. Res.*, 43:7566–7571, 2004.

[172] H.R. Null. Heat pumps in distillation. *Chem. Eng. Prog.*, 78:58–64, 1976.

[173] M. Nakaiwa, K. Huang, T. Ohmori, T. Akiya, and T. Takamatsu. Internally heat-integrated distillation columns: A review. *Trans. I. Chem. E*, 81A:162–177, 2003.

[174] Ullmann's. *Encyclopedia of Industrial Chemistry*. Germany, 5th. edition, 1995.

[175] J.M. Smith, H.C. Van Ness, and M.M. Abbott. *Introduction to Chemical Engineering Thermodynamics*. McGraw-Hill, 7th. ed., 2005.

[176] J. Gross and G. Sadowski. Application of the perturbed-chain saft equation of state to associating systems. *Ind. Eng. Chem. Res.*, 41(22):5510–5515, 2002.

[177] J. Gross and G. Sadowski. Modeling polymer systems using the perturbed-chain statistical associating fluid theory equation of state. *Ind. Eng. Chem. Res.*, 41(5):1084–1093, 2002.

[178] J. Gross. An equation-of-state contribution for polar components: Quadrupolar molecules. *AIChE J.*, 51(9):2556–2568, 2005.

[179] J. Gross and J. Vrabec. An equation-of-state contribution for polar components: Dipolar molecules. *AIChE J.*, 52(3):1194–1204, 2006.

[180] Sugata P. Tan, Hertanto Adidharma, and Maciej Radosz. Recent advances and applications of statistical associating fluid theory. *Ind. Eng. Chem. Res.*, 47(21):8063–8082, 2008.

[181] A. De Vos and B. Desoete. Equipartition Principles in Finite-Time Thermodynamics. *J. Non-Equilib. Thermodyn.*, 25:1–13, 2000.

[182] B. Andresen and J.M. Gordon. Constant thermodynamic speed for minimizing entropy production in thermodynamic processes and simulated annealing. *Phys. Rev. E*, 50:4346–4351, 1994.

# List of symbols

## Latin symbols

| | | |
|---|---|---|
| $A$ | $m^2$ | heat exchange area |
| $B$ | $mol.s^{-1}$ | bottom stream flow |
| $C_p$ | $J.K^{-1}$ | heat capacity at constant pressure |
| $C_{p,i}$ | $J.K^{-1}.mol^{-1}$ | molar heat capacity, constant pressure |
| $C_v$ | $J.K^{-1}$ | heat capacity at constant volume |
| $c_i$ | $mol.m^{-3}$ | concentration or molar density |
| $D$ | $m^2.s^{-1}$ | diffusion coefficient |
| $D_p$ | m | catalyst pellet diameter |
| $D_T$ | $m^2.s^{-1}.K^{-1}$ | thermal diffusion coefficient |
| D | $J.s^{-1}$ | the dissipation function, space integral |
| $D$ | m | diameter |
| $D$ | $mol.s^{-1}$ | distillate flow |
| $E$ | $V.m^{-1}$ | electric field |
| $E_{eq}$ | $V.m^{-1}$ | electric field of a system in equilibrium |
| $F$ | $C.mol^{-1}$ | Faraday's constant 96500 $C.mol^{-1}$ |
| $F$ | $mol.s^{-1}$ | feed flow |
| $F_i$ | $mol.s^{-1}$ | molar flow rate |
| $f$ | $Pa.s^{-1}$ | friction constant |
| $f_i$ | bar | fugacity |
| $G$ | J | Gibbs energy |
| $\Delta_r G$ | $J.mol^{-1}$ | reaction Gibbs energy |
| $g_i$ | $J.m^{-3}.mol^{-1}$ | partial molar Gibbs energy |
| $g$ | $m.s^{-2}$ | acceleration of gravity |
| $H$ | J | enthalpy |

| | | |
|---|---|---|
| $H_i$ | $J.mol^{-1}$ | partial molar enthalpy |
| $\mathcal{H}$ | $J.K^{-1}.s^{-1}$ | Hamiltonian in optimal control theory |
| | $J.K^{-1}.m^{-1}.s^{-1}$ | - 1D, stationary system |
| $\Delta_r H$ | $J.mol^{-1}$ | reaction enthalpy |
| $\Delta_{vap} H$ | $J.mol^{-1}$ | enthalpy of evaporation |
| $h$ | $J.m^{-3}$ | enthalpy per volume |
| $h$ | m | height |
| $j$ | $A.m^{-2}$ | electric current density |
| $j_{displ}$ | $A.m^{-2}$ | displacement current density |
| $J_q$ | $J.m^2.s^{-1}$ | flux of internal energy |
| $J_q'$ | $J.m^2.s^{-1}$ | flux of measurable heat |
| $J_i$ | $mol.m^2.s^{-1}$ | flux of component $i$ |
| $J_s$ | $J.K^{-1}.m^2.s^{-1}$ | flux of entropy |
| $k_B$ | $J.K^{-1}$ | Boltzmann's constant 1.381 $10^{-23} JK^{-1}$ |
| $K$ | | thermodynamic equilibrium constant |
| $L_{ik}$ | | phenomenological coefficient for coupling of fluxes i and k |
| $\mathcal{L}$ | $J.K^{-1}$ | Euler-Lagrange function |
| $L$ | m | system length |
| $L$ | $mol.s^{-1}$ | vapor flow |
| $l$ | m | length of box |
| $l_{ik}$ | | phenomenological coefficient for coupling of diffusional fluxes $i$ and $k$ |
| $m_i$ | kg | mass of component $i$ |
| $N_i$ | mol | amount of component $i$ |
| $N$ | mol | number of moles |
| $N$ | | number of trays |
| $n$ | | number of independent components |
| $P$ | $C\ m^{-2}$ | polarization density in the bulk phase |
| $p$ | bar (Pa) | pressure of the system |
| $p_{ext}$ | bar (Pa) | external pressure |
| $p_0$ | bar (Pa) | pressure of the surroundings |
| $p^{\ominus}$ | bar | standard pressure |
| $p_i^*$ | bar (Pa) | saturation pressure of pure gas $i$ |
| $Q_n$ | W | heat transferred on tray $n$ |

| | | |
|---|---|---|
| $q$ | J | heat delivered to the system |
| $q_0$ | J | heat delivered to the surroundings |
| $q^*$ | J.mol$^{-1}$ | heat of transfer |
| $R$ | J.K$^{-1}$.mol$^{-1}$ | gas constant 8.314 JK$^{-1}$ mol$^{-1}$) |
| $R_{ik}$ | | resistivity coefficient for coupling of fluxes $i$ and $k$ |
| $r_{ik}$ | | resistivity coefficient for coupling of diffusional fluxes $i$ and $k$ |
| $r$ | mol.m$^{-3}$.s$^{-1}$ | reaction rate |
| $S$ | J.K$^{-1}$ | entropy of the system |
| $s$ | J.K$^{-1}$.m$^3$ | entropy per volume |
| $S_0$ | J.K$^{-1}$ | entropy of the surroundings |
| $S^*$ | J.K$^{-1}$.mol$^{-1}$ | transported entropy (per mol) |
| $S_i$ | J.K$^{-1}$.mol$^{-1}$ | partial molar entropy |
| $S$ | J.K$^{-1}$.mol$^{-1}$ | molar entropy |
| $dS_{irr}/dt$ | J.K$^{-1}$.s$^{-1}$ | total entropy production |
| $\Delta_r S$ | J.K$^{-1}$.mol$^{-1}$ | reaction entropy |
| $T$ | K | absolute temperature |
| $T_0$ | K | temperature of the surroundings |
| $t$ | s | time |
| $t_{ion}$ | | transport number of ion |
| $t_i$ | | transference coefficient of $i$ |
| $U$ | J | internal energy |
| $U_i$ | J.mol$^{-1}$ | partial molar internal energy of $i$ |
| $u$ | J.m$^{-3}$ | internal energy per volume |
| $u_{ion}$ | m$^2$.s$^{-1}$.V$^{-1}$ | mobility of ion in electric field |
| $V$ | m$^3$ | volume |
| $V_i$ | m$^3$.mol$^{-1}$ | partial molar volume of $i$ |
| $v_i$ | m.s$^{-1}$ | particle velocity |
| $v$ | m.s$^{-1}$ | superficial gas velocity |
| $X_i$ | | general symbol, thermodynamic driving force |
| $w$ | J | work done on the system |
| $W$ | kg | weight |
| $x, y, z$ | m | coordinates |

| $x_i, y_i$ | | mole fraction of $i$ in liquid, and gas |
| $y_i$ | | activity coefficient for non-electrolyte |
| $z$ | $C.m^{-3}$ | charge density |

## Greek and mathematical symbols

| $\alpha$ | | transfer factor in the Butler-Volmer equation |
| $\delta$ | m | film thickness, surface thickness |
| $\epsilon$ | | catalyst bed porosity |
| $\phi$ | V | electric potential |
| $\varphi_i$ | | fugacity coefficient of $i$ in gas mixture |
| $\eta$ | V | overpotential |
| $\eta_C$ | | Carnot efficiency |
| $\eta_I$ | | first law efficiency |
| $\eta_{II}$ | | second law efficiency |
| $\gamma$ | | internal coordinate, mesoscopic description |
| $\gamma_i$ | | activity coefficient of $i$ in liquid mixture |
| $\Gamma$ | $mol.m^{-2}$ | adsorption |
| $\kappa$ | $S.m^{-1}$ | electrical conductivity |
| $\lambda$ | $W.K^{-1}.m^{-1}$ | thermal conductivity |
| $\lambda$ | | Lagrange multiplier function |
| $\mu_i$ | $J.mol^{-1}$ | chemical potential of $i$ |
| $\mu_{i,T}$ | $J.mol^{-1}$ | chemical potential of $i$ at constant $T$ |
| $\mu$ | $kg.m^{-1}.s^{-1}$ | viscosity |
| $\nu_{j,i}$ | | stoichiometric coefficient of component $i$ in reaction $j$ |
| $\xi$ | | degree of conversion |
| $\psi$ | V | Maxwell potential |
| $\pi$ | J | Peltier heat |
| $\Pi$ | | Poynting correction |
| $\rho$ | ohm.m | specific resistivity |
| $\rho$ | $kg.m^{-3}$ | density |
| $\rho_B$ | $kg.m^{-3}$ | catalyst bed density |
| $\sigma$ | $J.s^{-1}.K^{-1}.m^{-3}$ | local entropy production |

| $\tau$ | s | time lag |
|---|---|---|
| $\theta$ | | Heaviside function |
| $\Theta$ | s | process duration |
| $\Omega$ | m$^2$ | cross sectional area |
| $d$ | | differential |
| $\partial$ | | partial derivative |
| $\Delta$ | | change in a quantity |
| $\Sigma$ | | sum |
| $\equiv$ | | defined by |
| $\ominus$ | | denotes the standard state of 1 bar |
| inf | | denotes an infinitely dilute solution |

## Superscripts and subscripts

| | |
|---|---|
| a | super- or subscript meaning bulk anode |
| A | anion |
| a | subscript which means heating/cooling medium |
| B | super- or subscript meaning bottom stream |
| c | super- or subscript meaning cathode |
| c | subscript meaning cold fluid |
| C | cation |
| D | super- or subscript meaning distillate |
| eq | subscript meaning system in equilibrium |
| $e^-$ | property of electron |
| F | super- or subscript meaning feed stream |
| g | super- or subscript meaning gas phase |
| h | subscript meaning hot fluid |
| ig | superscript meaning ideal gas |
| l or L | super- or subscript meaning liquid phase |
| i | subscript meaning component i |
| r | subscript meaning reaction |
| s | superscript meaning surface |
| sat | superscript meaning saturated vapor phase |
| V | superscipt meaning vapor phase |
| $0i$ | subscript meaning pure component i |

# Index

Printed in the United States
By Bookmasters